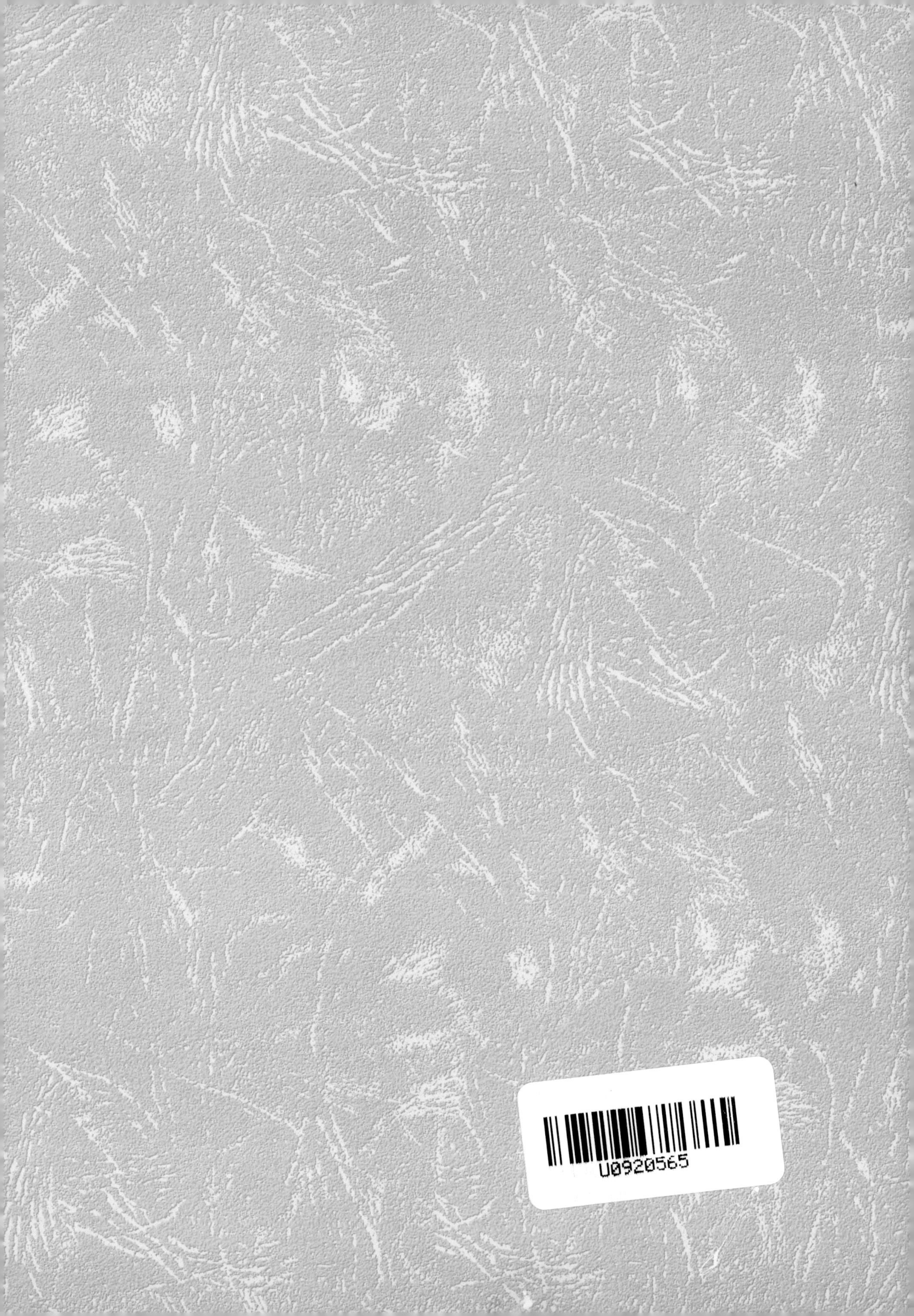

中国海洋统计年鉴

CHINA MARINE STATISTICAL YEARBOOK 2013

国 家 海 洋 局

Edited by
State Oceanic Administration,
People's Republic of China

海洋出版社

China Ocean Press

《中国海洋统计年鉴》编委会

周　忠	中国船舶工业集团公司
刘　悦	中国船舶重工集团公司
姜力孚	中国石油天然气集团公司
刘　岩	中国石油化工集团公司
金晓剑	中国海洋石油总公司
胡红江	中国盐业总公司
王华俊	中国有色金属工业协会
卢江宁	辽宁省海洋与渔业厅
王保民	河北省国土资源厅
孙连友	天津市海洋局
鲁小兵	山东省海洋与渔业厅
唐庆宁	江苏省海洋与渔业局
沈依云	上海市海洋局
童加朝	浙江省海洋与渔业局
黄世峰	福建省海洋与渔业厅
洪伟东	广东省海洋与渔业局
刘　斌	广西壮族自治区海洋局
潘建纲	海南省海洋与渔业厅
栾玉瑄	大连市海洋与渔业局
于成撲	青岛市海洋与渔业局
吴建义	宁波市海洋与渔业局
王春生	厦门市海洋与渔业局
梁俊乾	深圳市海洋局

《中国海洋统计年鉴》编辑部

主　任：

何广顺　　国家海洋信息中心副主任

副主任：

沈　君　　国家海洋局政策法规和规划司海洋经济处处长

王晓惠　　国家海洋信息中心海洋经济部主任

成　员：

王占坤　李长如　董　伟　杨　娜　郭　越　周洪军

林香红　蔡大浩　郑　莉　宋维玲　李巧稚　张潇娴

赵心宇　张　伟　朱　凌　徐丛春　李宜良

（国家海洋信息中心）

曹　斐（国家海洋局政策法规和规划司海洋经济处）

特邀编辑：

李燕丽　教育部

林　涛　科技部

苗　强　国土资源部

陈　钟　交通运输部

张淑娜　水利部

郭　睿　农业部

毛玉如　环境保护部

刘建杰　国家林业局

周　鲲　国家旅游局

崔胜先　中国科学院

张淑丽　中国地震局

骆　婷　中国气象局

Editorial Board of
China Marine Statistical Yearbook

Shi Bing	Chinese Academy of Sciences
Han Zhiqiang	China Earthquake Administration
Xie Pu	China Meteorological Administration
Zhou Zhong	China State Shipbuilding Corporation
Liu Yue	China Shipbuilding Industry Corporation
Jiang Lifu	China National Petroleum Corporation
Liu Yan	China Petrochemical Group Corporation
Jin Xiaojian	China National Offshore Oil Corporation
Hu Hongjiang	China National Salt Industry Corporation
Wang Huajun	China Nonferrous Metals Industry Association
Lu Jiangning	Department of Ocean and Fisheries of Liaoning Province
Wang Baomin	Hebei Province Department of Land and Resources
Sun Lianyou	Tianjin Ocean Administration
Lu Xiaobing	Department of Oceanic and Fishery of Shandong Province
Tang Qingning	Jiangsu Provincial Ocean and Fisheries Bureau
Shen Yiyun	Shanghai Ocean Administration
Tong Jiachao	Zhejiang Provincial Ocean and Fisheries Bureau
Huang Shifeng	Department of Oceanic and Fishery of Fujian Province
Hong Weidong	Guangdong Provincial Oceanic and Fishery Administration
Liu Bin	The Oceanic Administration of Guangxi
Pan Jiangang	Department of Marine and Fishery of Hainan Province
Luan Yuxuan	Dalian Ocean and Fishery Administration
Yu Chenpu	Qingdao Ocean and Fishery Administration
Wu Jianyi	Ningbo Ocean and Fishery Bureau
Wang Chunsheng	Oceans and Fisheries Bureau of Xiamen
Liang Junqian	Shenzhen Ocean Administration

Editorial Department of China Marine Statistical Yearbook

Luo Ting	China Meteorological Administration
Qiu Yinlang	China State Shipbuilding Corporation
Xu Jing	China Shipbuilding Industry Corporation
Sun Xiaoxian	China National Petroleum Corporation
Sun Yan	China Petrochemical Group Corporation
Hao Jinghui	China National Offshore Oil Corporation
Du Lihua	China National Salt Industry Corporation
Sun Xiumin	China Nonferrous Metals Industry Association
Wei Nan	Department of Ocean and Fisheries of Liaoning Province
Huang Miao	Hebei Province Department of Land and Resources
Yang Jinlu	Tianjin Ocean Administration
Shu Qingmeng	Department of Oceanic and Fishery of Shandong Province
Pei Pei	Jiangsu Provincial Ocean and Fisheries Bureau
Zhang Cheng	Shanghai Municipal Maritime Affairs Administration
Ying Xiao	Zhejiang Provincial Ocean and Fisheries Bureau
Liao Runxue	Department of Oceanic and Fishery of Fujian Province
Li Hongbing	Guangdong Provincial Oceanic and Fishery Administration
Chen Heng	The Oceanic Administration of Guangxi
Lin Yun	Department of Marine and Fishery of Hainan Province
Shen Ronglian	Dalian Ocean and Fishery Administration
Wang Zijiang	Qingdao Ocean and Fishery Administration
Wu Mingchun	Ningbo Ocean and Fishery Bureau
Chi Xincai	Oceans and Fisheries Bureau of Xiamen
Luo Xiaoxia	Shenzhen Municipal Economic, Trade and Informanization Commission

Executive Editor: Li Changru

English Proof-reader: Lin Baofa

编 者 说 明

一、《中国海洋统计年鉴》2013 年版是一部全面反映 2012 年中华人民共和国海洋经济发展和海洋管理服务情况的资料性年鉴，全书为中英文对照。

二、本年鉴的统计资料范围为人们在海洋和沿海地区开发、管理、利用海洋资源和空间，发展海洋经济的生产和活动以及沿海地区的社会经济概况。地域范围为沿海地区、沿海城市和沿海地带，其排列顺序按《沿海行政区域分类与代码》(HY/T 094-2006)的顺序排列。

三、本年鉴内容包括综合资料、海洋经济核算、主要海洋产业活动、主要海洋产业生产能力、涉海就业、海洋科学技术、海洋教育、海洋环境保护、海洋行政管理及公益服务、全国及沿海社会经济、部分世界海洋经济统计资料等十一部分。

四、本年鉴根据《海洋统计报表制度》（国统制[2012]50 号）和《海洋生产总值核算制度》（国统制[2013]93 号），资料主要来源于沿海省、自治区、直辖市统计局、海洋厅（局）以及 20 个有关涉海部、局、总公司。

五、本年鉴中除特殊说明外，所有价值量指标均为当年价，除注明年份外，其他均为 2012 年数据，年鉴中每部分后附有主要海洋统计指标解释，对主要海洋统计指标的含义、统计范围和统计方法做了简要说明。统计数据中的其他说明置于表的下面。有续表的资料，如有注释均置于最后一张表的下面。

六、本年鉴表格中符号使用说明：“…”表示数据不足本表最小计算单位数；“空格”表示该项统计指标数据不详或无该项数据；“#”表示其中的主要项；其他符号如“*”或“①”等表示本表后面有注释。

七、本年鉴中由于数字精确度的原因，四舍五入后部分分项值之和与合计值有微小差异。

八、本年鉴资料国内部分未包括香港特别行政区、澳门特别行政区和台湾省数据。

九、《中国海洋统计年鉴》在编撰过程中，得到了各有关单位的大力支持，我们在此表示衷心的感谢。本年鉴中如有疏漏和不妥之处，敬请读者批评指正。

《中国海洋统计年鉴》编辑部

Introduction

I. *China Marine Statistical Yearbook* (2013) is a data almanac which reflects in an all-round way the development of marine economy, marine management and service in the People's Republic of China in 2012, and it is a Chinese-English bilingual edition.

II. The Yearbook's statistics cover the production and activities in the marine and coastal areas in relation to the development, management and utilization of marine resources and space, and the development of marine socioeconomy. The regions covered are the coastal regions, coastal cities and coastal zones with coastlines, which are arranged in order according to the *Coastal Administrative Areas Classification and Codes* (HY/T 094—2006).

III. The data in the yearbook consist of 11 sections, namely, integrated data, marine economic accounting, major marine industrial activities, production capacity of major marine industries, ocean-related employment, marine science and technology, marine education, marine environmental protection, marine administration and public-good service, national and coastal socioeconomy, part of the world's marine economic statistics data.

IV. The Yearbook is based on the *Marine Statistics Report System* (Guotongzhi [2012] No. 50) and the *Ocean Gross Product Accounting System* (Guotongzhi [2013] No. 93), its data mainly come from the statistical bureaus and the oceanic administrations of the coastal provinces, autonomous regions, and municipalities directly under the Central Government as well as the 20 ocean-related ministries, bureaus and general corporations concerned.

V. Unless otherwise specified in the Yearbook, all the value indicators are given at the current price. Each section is attached by explanatory notes to the major marine statistical indicators, giving a brief explanation for the meaning, statistical range and statistical methods of the major marine statistical indicators. Other notes to the statistical data are listed below the tables. For the data with continued tables, annotations, if any, are put below the last table.

VI. The usage of symbols in the tables: "…" indicates the statistics smaller than the minimum calculation unit in the table; "Blank" indicates that the data of the statistical index is unknown for the time being or that there shouldn't be any data; "#" indicates the major items of the table; Other symbols, such as "*" or "①", indicate "see footnotes below".

VII. For reasons of digital accuracy, there is small difference between the sum of values of some subterms after having been rounded off and the total values.

VIII. The domestic part of the Yearbook does not include that of Hong Kong Special Administrative Region, Macau Special Administrative Region and Taiwan Province.

IX. In the course of editing *China Marine Statistical Yearbook*, we enjoyed energetic support from the various departments concerned and we hereby extend our heartfelt thanks to them. Criticisms and comments are welcome from readers on any of the oversights and inappropriateness as the time for editing is too short.

Editorial Department of
China Marine Statistical Yearbook

目　次

CONTENTS

3 主要海洋产业活动
Major Marine Industrial Activities

4 主要海洋产业生产能力

Production Capacity of Major Marine Industries

9 海洋行政管理及公益服务
Marine Administration and Public-Good Service

10 全国及沿海社会经济
National and Coastal Socioeconomy

11 部分世界海洋经济统计资料

Part of the World's Marine Economic Statistics Data

2012年我国海洋经济发展综述

2012年，沿海各地区认真落实党中央、国务院发展海洋经济的战略部署，以科学发展为主题，加快推进经济发展方式转变，海洋经济继续保持平稳增长的良好势头。

一、全国海洋经济发展概况

2012年，全国海洋生产总值50 045.2亿元，比上年增长8.06%（除特别注明外，增长率均按可比价计算），海洋生产总值占国内生产总值的9.64%，占沿海地区生产总值的15.84%。全国涉海就业人员3 468.8万人，比上年增加47.1万人。

二、主要海洋产业发展情况

2012年主要海洋产业实现增加值20 829.9亿元，比上年增长7.0%，占海洋生产总值的41.62%，滨海旅游业和海洋交通运输业仍占主导地位。

海洋第一产业 2012年，海洋渔业快速发展，海洋水产品产量稳步增长。海水产品产量3 033.3万吨，比上年增长4.3%，其中，海洋捕捞产量1 267.2万吨，比上年增长2.0%；海水养殖面积218.1万公顷，比上年略有增长，海水养殖产量1 643.8万吨，比上年增长6.0%，远洋渔业产量122.3万吨，比上年增长6.6%。此外，沿海地区中心渔港达60个，一级渔港77个，比上年增加10个。

海洋第二产业 2012年，海洋第二产业保持平稳的发展态势。受蓬莱19-3油田停产，以及产能控制、台风海冰的影响，海洋油气业增速回落，全年实现增加值1 718.7亿元，比2011年增长0.2%，全年海洋原油产量4 444.8万吨，比2011年减少0.2%，海洋天然气产量122.8亿立方米，比2011年增长1.1％。海洋矿业实现增加值45.1亿元，比2011年下降13.5%。受不利天气及盐田面积减少等因素影响，我国海盐出现减产情况，2012年海洋盐业实现增加值60.1亿元，比上年减少24.3%，海盐产量为2 986.4万吨，比上年减少10.1%。海洋化工产业实现增加值843.0亿元，比上年增长26.2%。随着国家对海洋生物医药业政策扶持和投入力度的逐步加大，海洋生物医药业发展势头良好，全年实现增加值184.7亿元，比2011年增长22.3%。2012年多个海上风电场的相继投产，海洋电力业发展势头总体良好，全年实现增加值77.3亿元，比2011年增长25.9%。我国海水利用产业呈现稳步发展态势，全年实现增加值11.1亿元，比2011年增长3.8%。海洋船舶工业受全球航运市场持续低迷的影响，交船难、接单难、盈利难等问题突出，全年实现增加值1 291.3亿元，比2011年减少4.0%。海洋工程建筑业继续保持增长态势，新开工项目和在建工程稳步推进，全年实现增加值1 353.8亿元，比上年增长22.6%。

海洋第三产业　2012年，全国沿海港口生产态势总体良好，但受国内外宏观经济环境影响，增速持续放缓，全年实现增加值4 752.6亿元，比上年增长5.2%。沿海港口货物吞吐量687 975万吨，比上年增长8.2%，国际标准集装箱吞吐量15 797万标准箱，比上年增长8.0%。滨海旅游产业规模持续扩大，沿海地区积极推出优秀滨海旅游开发项目，吸引各类企业投资，滨海旅游内容不断充实、主题不断丰富、线路不断增加、投资持续升温，滨海旅游产品体系日益呈现多元化发展。滨海旅游业全年实现增加值6 931.8亿元，比2011年增长8.9％。

三、区域海洋经济发展情况

2012年，区域海洋经济总量平稳增长，但受国际严峻形势的影响和国家“转方式、调结构”的总体部署，大部分沿海地区海洋经济增速有所放缓。环渤海经济区、长江三角洲经济区和珠江三角洲经济区海洋生产总值分别为17 925.1亿元、15 616.7亿元和10 506.6亿元，占全国海洋生产总值的比重分别为35.8%、31.2%、21.0%。

环渤海经济区海洋经济增速放缓，海洋生产总值比上年增长9.7%（现价），占地区生产总值比重15.7%。海洋产业增加值为10 428.3亿元，海洋相关产业增加值为7 496.8亿元。海洋渔业、海洋油气业、海洋交通运输业、滨海旅游业作为环渤海经济区四大海洋支柱产业，增加值合计达到7011.0亿元，占该地区主要海洋产业增加值的84.2%。海洋生物医药业、海洋电力业、海水利用业等海洋新兴产业增幅较大，分别比上年增长22.7%、35.6%、22.1%（现价）。

长江三角洲经济区增速持续减缓，海洋生产总值比上年增长8.4%（现价），海洋生产总值占地区生产总值比重14.3%。长江三角洲海洋产业增加值8 966.2亿元，海洋相关产业增加值6 650.6亿元。按照产值贡献，滨海旅游业、海洋交通运输业、海洋船舶工业和海洋渔业四个产业位居前列，其增加值之和占该地区主要海洋产业增加值的91.5%，其中滨海旅游业占该地区主要海洋产业增加值的41.1%，产值贡献位居第一。海洋生物医药业、海水利用业、海洋油气业、海洋电力业增速较快，分别比上年增长31.8%、24.5%、21.8%和20.3%（现价）。

珠江三角洲经济区海洋经济稳步增长，在地区经济和全国海洋经济发展中发挥着重要作用。海洋生产总值达10 506.6亿元，比上年增长7.2%（现价），海洋生产总值占地区生产总值比重达18.4%。海洋产业增加值为6 317.0亿元，海洋相关产业增加值为4 189.6亿元。滨海旅游业、海洋交通运输业、海洋油气业、海洋化工业和海洋渔业依然为珠江三角洲经济区海洋经济发展的支柱产业，其增加值之和占该地区主要海洋产业增加值90.2%。海洋工程建筑业和海洋电力业增长迅速，增长速度分别达到59.7%和29.9%（现价）。

四、海洋科研教育

2012年，海洋科研教育事业继续保持稳步发展，统计的海洋科研机构共177个，从业人员37 679人，比2011年增长0.6%；海洋科研机构承担海洋科技课题15 403项，比2011年增长8.1%；发表海洋科技论文16 713篇，出版海洋科技著作338种，分别比2011年增长7.5%和21.6%；专利授权数2 746件，其中发明专利1 883件，分别比2011年增长35.0%和39.0%。开设海洋专业的高等院校达361个，比上年增长4.3%，专任教师271 996人，比上年增长7.1%。高等教育和中等职业教育海洋专业毕业生数106 120人，比上年增长8.9%；招生人数和在校人数分别为83 353人和272 323人，分别比上年下降16.9%和11.8%。

五、海洋环境保护

2012年，我国海洋环境质量状况总体维持在较好水平。符合第一类海水水质标准的海域面积约占我国管辖海域面积的94%，海洋沉积物质量良好。海水、海洋沉积物、海洋生物的放射性水平和海洋大气γ辐射空气吸收剂量率均处于本底范围内。国家级海洋保护区环境质量总体良好，主要保护对象或保护目标基本保持稳定。重点海水浴场、滨海旅游度假区环境质量总体良好。海水增养殖区环境质量基本满足养殖活动要求。海洋倾倒区环境状况总体稳定，未因倾倒活动产生明显影响。但是，近岸海域水体污染、生态受损、灾害多发等环境问题依然突出。陆源入海排污口达标排放率依然较低，约1.9万平方公里海域呈重度富营养化状态；81%实施监测的河口、海湾等典型海洋生态系统处于亚健康和不健康状态；发生赤潮的累计面积较上年有所增加；渤海滨海平原地区海水入侵和土壤盐渍化严重；局部地区侵蚀速度加快。2012年，我国共发生138次风暴潮、海浪和赤潮过程，各类海洋灾害（含海冰、绿潮等）造成直接经济损失155.25亿元，死亡（含失踪）68人。

六、海洋行政管理

2012年，海洋综合管控能力继续加强，行政管理工作稳步推进。全年颁发海域使用权证书2 348本，比2011年减少39.4%；确权海域面积27.17万公顷，比2011年增长46.1%；发放海岛使用权证书5本，确权海岛面积30.28公顷，确权海岛5个。全年共签发疏浚物海洋倾倒许可证146份，实施各项海洋执法检查共73 311次，发现违法行为1 443起，比2011年减少31.7%；提供海洋数值预报服务共36 202次，开展海洋调查项目427个，获得数据共22.58万个，各项海洋观测获得数据量共28 475.59万个，全年接收存档卫星遥感数据量共计31 648.50 GB；本年接收纸质档案18 214卷（册），电子档案9 898 GB；共有53项行业标准通过立项审查，发布行业标准1项。

Summary of China′s Marine Economic Development in 2012

In 2012, all coastal regions conscientiously put into effect the strategic plan of the Central Party Committee and the State Council for the development of marine economy and, with the scientific development as the key subject, speed up the advancement of the transformation of the pattern of economic development so that the marine economy continues to keep a good momentum of stable growth.

I. Survey of the National Marine Economic Development

In 2012, the national gross ocean product amounts to 5 004.52 billion yuan (RMB), 8.06% up from the previous year (Unless otherwise specified, the growth rate is calculated at the comparable price), accounting for 9.64% of the GDP and 15.84% of the Gross Regional Product. The number of people employed by the ocean related sectors throughout the country reaches 34.688 million, 0.471 million more than that in the previous year.

II. Development of Major Marine Industries

The major marine industries realize an added value of 2 082.99 billion yuan in 2012, growing by 7.0% as compared with that of last year, accounting for 41.62% of the gross ocean product, and the coastal tourism and marine communications and transportation still occupy the leading position.

Primary marine industry

In 2012, marine fishery develops rapidly and the output of marine aquatic products grows steadily. The production of marine aquatic products amounts to 30.333 million tons, 4.3% up from the previous year, among which, the yield from marine fishing is 12.672 million tons, 2.0% up from the previous year; the area of mariculture 2.181 million hm^2, slightly up from the previous year; the output of mariculture 16.438 million tons, 6.0% up from the previous year and that of deep-sea fishing 1.223 million tons, 6.6% up from the previous year. Moreover, the number of central fishing ports in the coastal region reaches 60, and that of first-class fishing ports 77, 10 more than that in the previous year.

Secondary marine industry

In 2012, the secondary marine industry keeps a smooth and stable posture of development. Affected by the stop of production of the Penglai 19-3 Oilfield as well as production capacity control, typhoon and sea ice, the growth rate of the offshore oil and gas industry has fallen, a full-year value added of 171.87 billion yuan is effected, 0.2% up from 2011, the output of marine crude oil for the whole year is 44.448 million tons, 0.2% down from 2011, and the production of marine natural gas 12.28 billion m^3, growing by 1.1% as against that in 2011. The marine mining industry effects an added value of 4.51 billion yuan, 13.5% down from 2011. Influenced by such factors as unfavorable weather and the decrease of salt-pan area etc., the production of sea-salt industry in China has dropped, the marine salt industry has effected a value added of 6.01 billion yuan, 24.3% down from the previous year and the production of sea salt is 29.864 million tons, decreasing by 10.1% as compared with that in the previous year. The marine chemical industry has effected an added value of 84.30 billion yuan, 26.2% up from the previous year. With the gradual increase of the state's policy support and investment effort for the marine biomedicine industry, the marine biomedicine industry keeps a good momentum of development, realizing a full-year added value of 18.47 billion yuan, 22.3% up from 2011. In 2012, a number of offshore wind farms have been put into production successively, and the development momentum of the marine electric power industry is good in general, effecting a full-year value added of 7.73 billion yuan, 25.9% up from 2011. The seawater utilization industry of our country presents a posture of steady development, realizing a full-year value added of 1.11 billion yuan, 3.8% up from 2011. Affected by the sustained downturn of the global shipping market, the marine shipbuilding industry encounters the outstanding difficulty in delivering ships, receiving orders and making profit, etc., effecting a full-year added value of 129.13 billion yuan, 4.0% down from 2011. The marine engineering construction industry continues to keep a posture of growth, the newly started projects and the engineering under construction are being carried forward steadily, and the industry realizes a full-year added value of 135.38 billion yuan, 22.6% up from the previous year.

Tertiary marine industry

In 2012, the production situation of national coastal harbours is on the whole good, but affected by the micro economic environment both at home and abroad, their production growth rate continues to slow down, effecting a full-year added value of 475.26 billion yuan, 5.2% up from the previous year. The cargo handling capacity of coastal harbours is 6.879 75 billion tons, 8.2% up from the previous year and the handling capacity of international standardized containers 157.97 million standard cases, 8.0% up from the previous year. The scale of the coastal tourism industry continues to expand and the coastal regions actively put out fine projects of coastal tourism so as to attract the investment by various enterprises. The content of coastal tourism is constantly substantiated, the themes are constantly enriched, the lines of tourism increase and the investment continues to be warming. And the series of coastal tourist products are increasingly presenting a diversified development. The coastal tourism accomplishes a full-year added value of 693.18 billion yuan, 8.9% up from 2011.

III. Development of Regional Marine Economy

In 2012, the regional marine economic aggregate grows steadily, but subject to the influence of the severe international situation and the national overall plan of changing the mode and adjusting the structure, the growth rate of marine economy in most of the coastal region has somewhat slowed down. The gross ocean products of the Round-the-Bohai Economic Zone, Changjiang River Delta Economic Zone and Zhujiang River Delta Economic Zone are 1 792.51 billion yuan, 1 561.67 billion yuan and 1 050.66 billion yuan respectively, accounting for 35.8%, 31.2% and 21.0% of the national gross ocean product separately.

The marine economic growth rate of the Round-the-Bohai Economic Zone has slowed down, and the gross ocean product has increased by 9.7% as compared with that in the previous year (at the current price), occupying 15.7% of the gross regional product. The value added of marine industries is 1 042.83 billion yuan and that of the marine related industries 749.68 billion yuan. The added values of marine fishery, offshore oil and gas industry, marine communications and transportation and coastal tourism as the four pillar industries of the Round-the-Bohai Economic Zone amount to 701.10 billion yuan, accounting for 84.2% of the added value of major marine industries in the region. The growth rates of the emerging marine industries such as marine biomedicine, marine electric power industry and seawater utilization industry increase by a larger margin, 22.7%, 35.6% and 22.1% up from the previous year respectively (at the current price).

The growth rate of the Changjiang River Delta Economic Zone continues to slow down and its gross ocean product increases by 8.4% (at the current price) as against that in the previous year and accounts for 14.3% of the gross regional product. The value added of the marine industries in the Changjiang River Delta is 896.62 billion yuan and that of the marine related industries 665.06 billion yuan. In the light of the output value contributions, the industries of coastal tourism, marine communications and transportation, marine shipbuilding and marine fishery rank among the first and the sum total of their added values accounts for 91.5% of the added value of major marine industries in the region, among which, the added value of coastal tourism occupies 41.1% of the added value of major marine industries in the region and its output value contribution ranks first. The growth rates of the industries of marine biomedicine, seawater utilization, offshore oil and gas, and marine electric power are faster, 31.8%, 24.5% 21.8% and 20.3% up from the previous year separately (at the current price).

Marine economy in the Zhujiang River Delta Economic Zone is steadily growing and playing an important role in the development of the regional economy and the national marine economy. The gross ocean product of this zone reaches 1 050.66 billion yuan, 7.2% up from the previous year (at the current price), and the gross ocean product occupies a proportion of 18.4% of the gross regional product. The value added of marine industries is 631.70 billion yuan and that of marine related industries 418.96 billion yuan. Coastal tourism, marine communications and transportation, offshore oil and gas industry, marine chemical industry and marine fishery are still the pillar industries in the marine economic development of the Zhujiang River Delta Economic Zone, the sum total of their values added accounting for 90.2% of the added values of major marine industries in the region. The marine engineering construction industry and the marine electric power industry grow rapidly, their growth rates reaching 59.7% and 29.9% respectively (at the current price).

IV. Marine Scientific Research and Education

In 2012, the marine scientific research and education continue to develop steadily. The statistical number of marine scientific research institutions totals 177 with 37 679 employees, increasing by 0.6% as compared with 2011; the number of marine scientific and technological projects undertaken by these institutions is 15 403, 8.1% up from 2011; 16 713 marine scientific and technological papers and 338 kinds of marine scientific and technological works are published, 7.5% and 21.6% up from 2011 separately; the number of patents authorized is 2 746, of which, 1 883 is the patents for discovery, increasing by 35.0% and 39.0% separately as against those in 2011. The number of the institutions of higher learning which offer marine specialties amounts to 361, 4.3% up from 2011, and the number of full-time teachers in this regard is 271 966, 7.1% up from last year; the number of graduates from the marine speciality of higher learning and secondary vocational education is 106 120, 8.9% up from 2011; and the number of students enrolled and that in school are 83 353 and 272 323 respectively, 16.9% and 11.8% down from the previous year separately.

V. Marine Environmental Protection

In 2012, China's marine environmental condition is on the whole maintained at a good level. The area of the sea that is up to the standard of first-class seawater quality accounts for about 94% of the sea area under China's jurisdiction, and the quality of marine sediment is fine. Both the radioactivity level of seawater, marine sediment and marine life, and the rate of air absorbed dose of the marine atmospheric γ radiation are within the background range. The environmental quality of the national marine protected area (MPA) is good on the whole and the marine objects or targets of protection basically keep stable. The environmental quality in the key bathing beaches and costal tourist resorts is good on the whole. The environmental quality in the seawater culture and stock enhancement zone can basically satisfy the demand for cultivating activities. The environmental condition of the ocean dumping zone is stable on the whole and has not been significantly affected by dumping activities. However, the environmental problems in the nearshore sea area such as water pollution, biological damage, frequent occurrence of disasters etc. are still prominent. The standard discharge rate of the terrigenous sewage discharge exits is still lower and about 19 000 km^2 sea area is in a state of serious eutrophication; 81% of the typical marine ecosystem that has been monitored such as estuaries, bays etc. is in a sub-healthy or unhealthy state; the total area where red tides occur has somewhat increased as compared with that in the previous year; in the coastal plain area of the Bohai Sea, seawater intrusion and soil salinization are serious; and in local area, erosion has speeded up. In 2012, a total of 138 storm surge, sea wave and red tide processes take place in China, the direct economic loss caused by various marine disasters (containing sea ice, green tide, etc.) reaches 15.525 billion yuan, which result in the death (missing) of 68 people.

VI. Marine Administration

In 2012, the integrated capacity of marine management and control continues to be improved and the administrative management has been carried forward steadily. A total of 2 348 certificates for the sea area use right are issued for the whole year, 39.4% down from 2011; the sea area with the ownership of patent rights reaches 271 700 hm^2, 46.1% up from 2011; 5 certificates for the sea island use right are issued, and the area and number of sea island with the ownership of patent rights reaches 30.28 hm^2 and a total of 146 permits for oceanic dumping of dredged material are signed and issued throughout the year, and marine inspections for law enforcement are carried out on 73 311 occasions, during which 1 443 cases of unlawful practice are discovered, 31.7% down from 2011; the marine numerical forecast service is provided for 36 078 times, 427 marine survey projects are carried out, acquiring 225 800 data and the data quantity from various marine observations totals 284.755 9 million and the quantity of satellite remote-sensing data received and placed on file for the whole year totals 31 648.50GB; the paper archives received this year total 18 214 volumes (copies) and the electronic archives 9 898GB; and a total of 53 items of professional standard have been authorized and examined, and one item of professional standard has been issued.

图 1 全国海洋生产总值及三次产业构成

Chinas Gross Ocean Product and Three Industries Composition

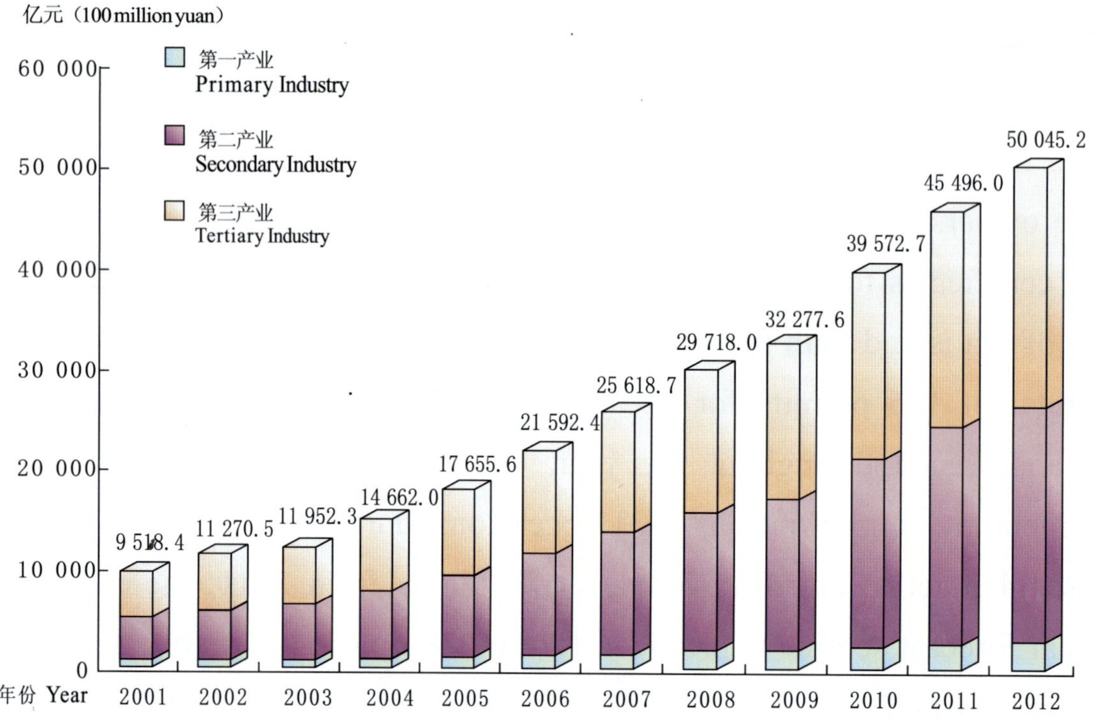

图 2 2012年全国主要海洋产业增加值构成

Composition of Added Values of the Major Marine Industries in China in 2012

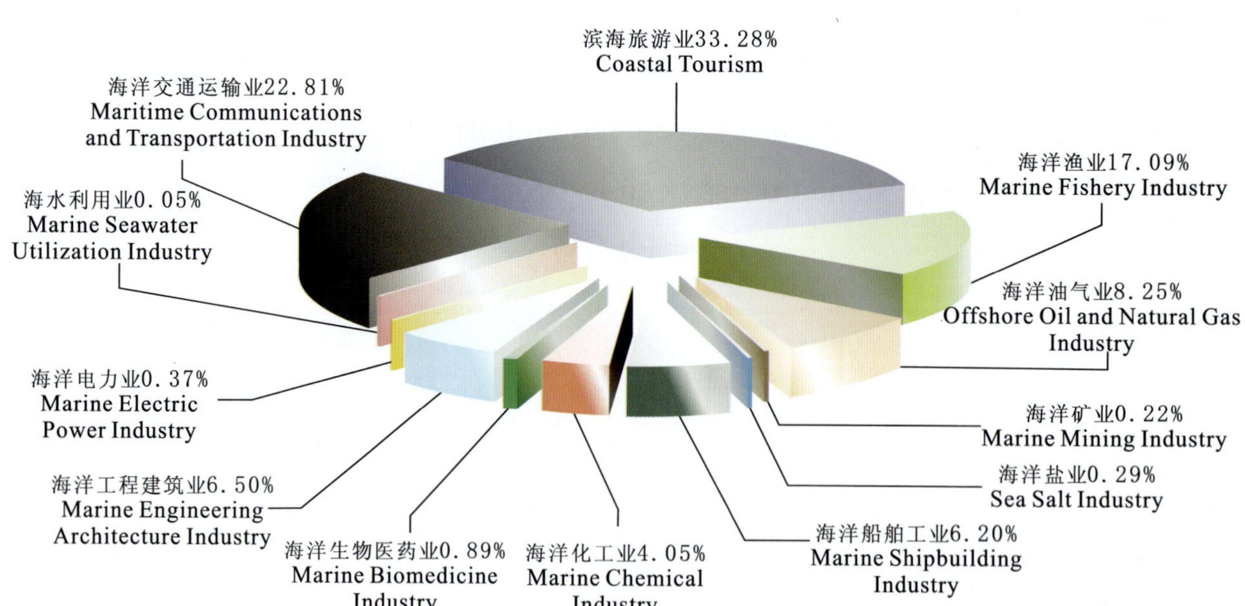

图 3 2012年沿海地区海洋生产总值

Gross Ocean Product by Coastal Regions in 2012

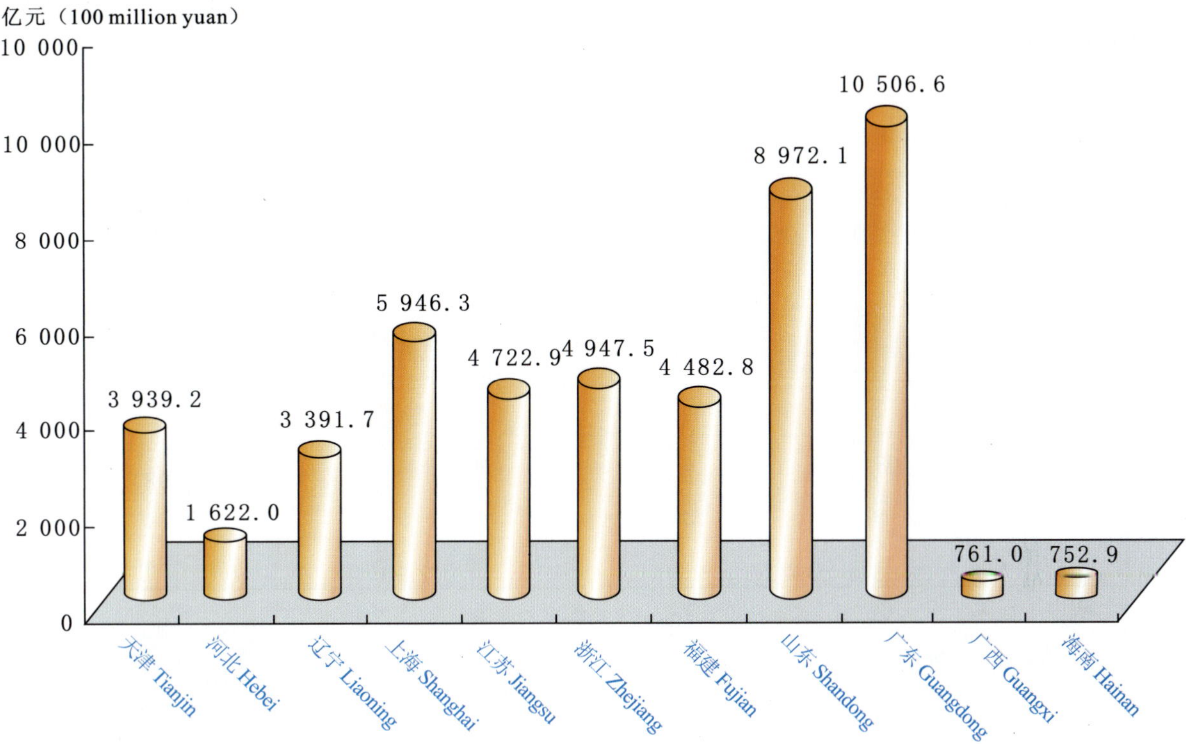

图 4 全国海洋捕捞养殖产量

National Marine Catches and Mariculture Production

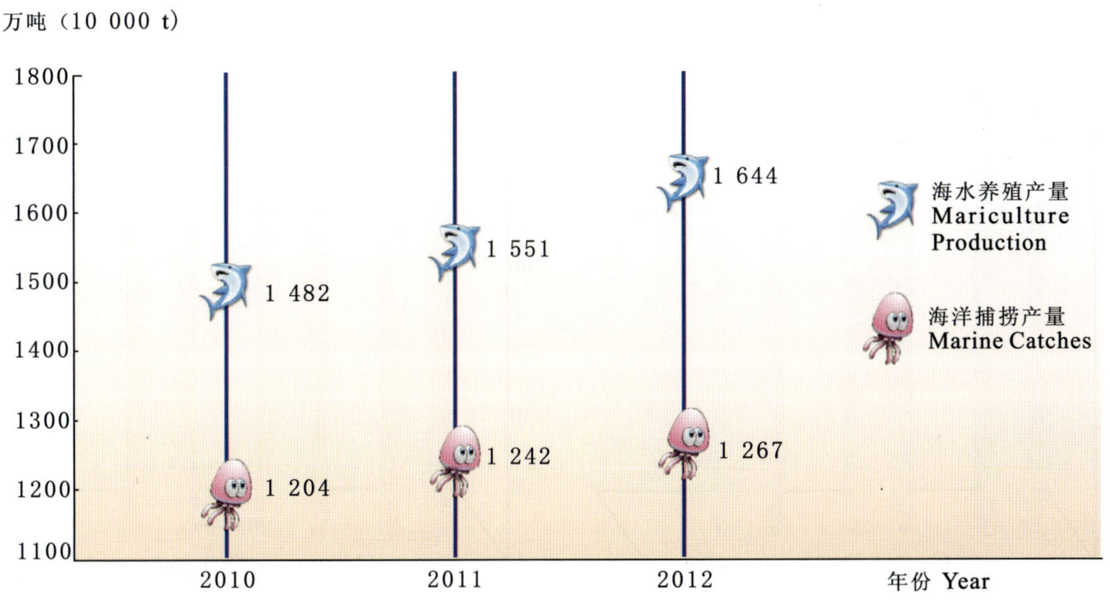

图 5　2012年沿海地区海洋捕捞养殖产量

Marine Catches and Mariculture Production by Coastal Regions in 2012

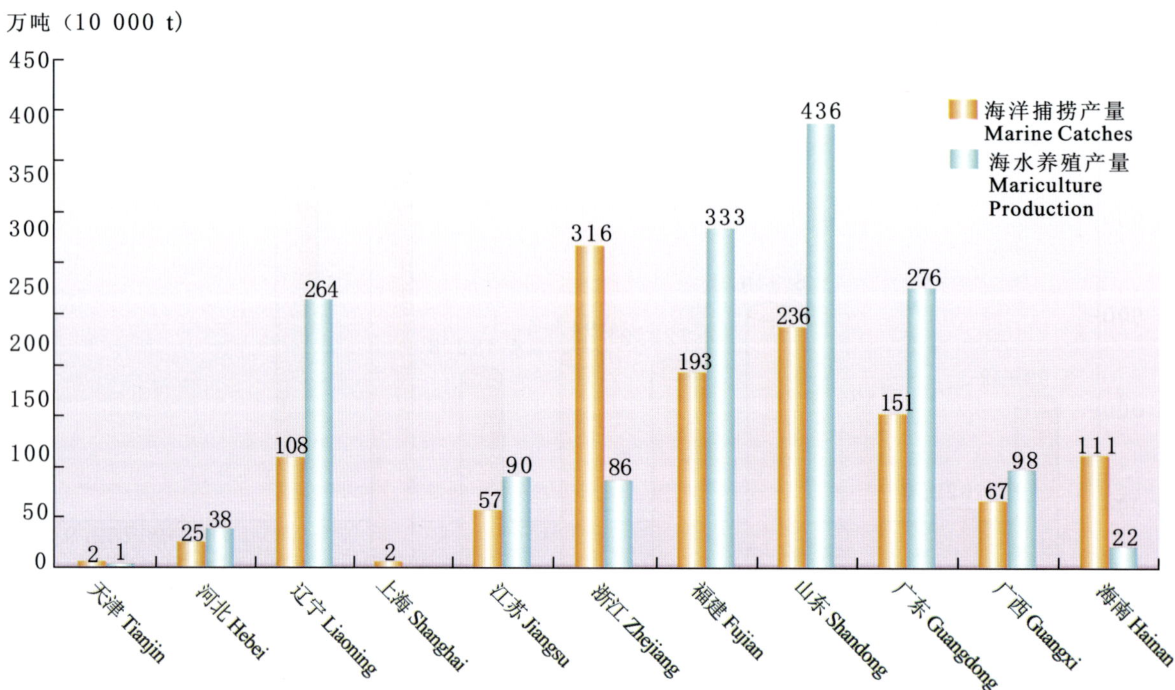

图 6　全国海洋石油和天然气产量

National Output of Offshore Oil and Natural Gas

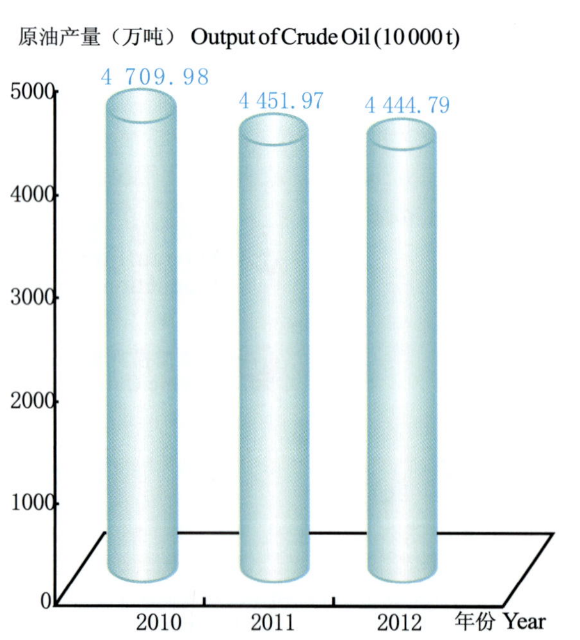

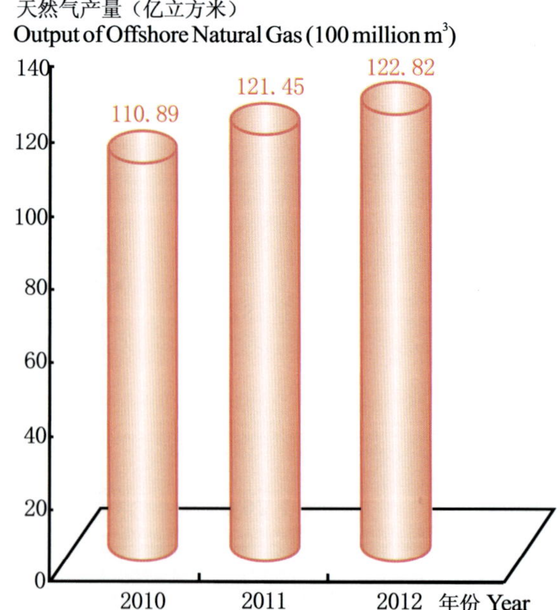

图 7 海洋原油产量占全国原油产量比重
Proportion of Offshore Crude Oil Production in the National Total

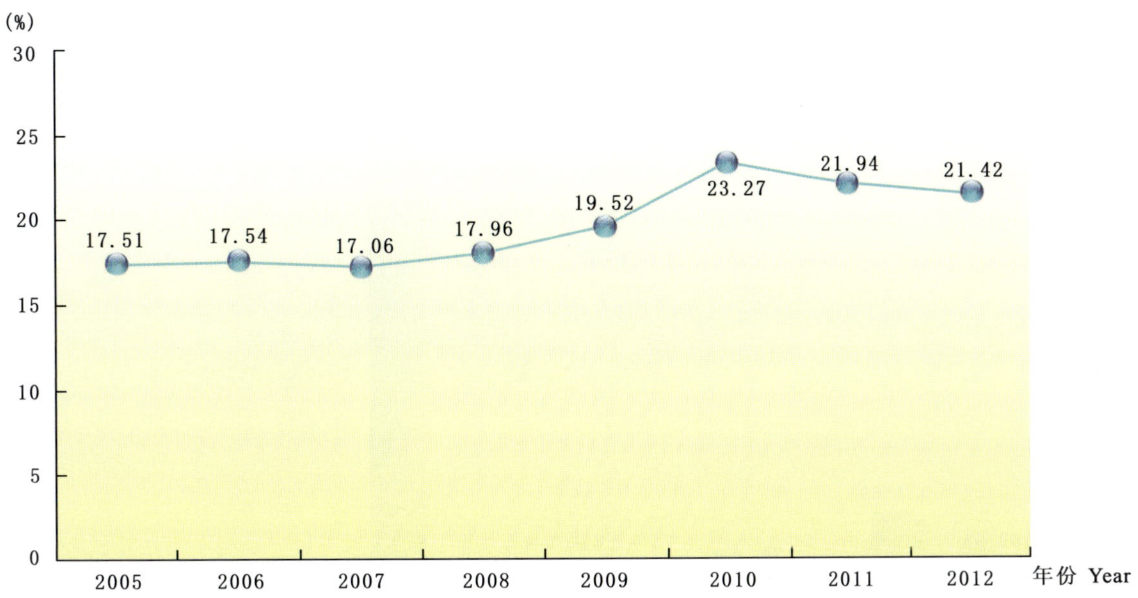

图 8 2012年海洋矿业产量
Production of Marine Mining Industry in 2012

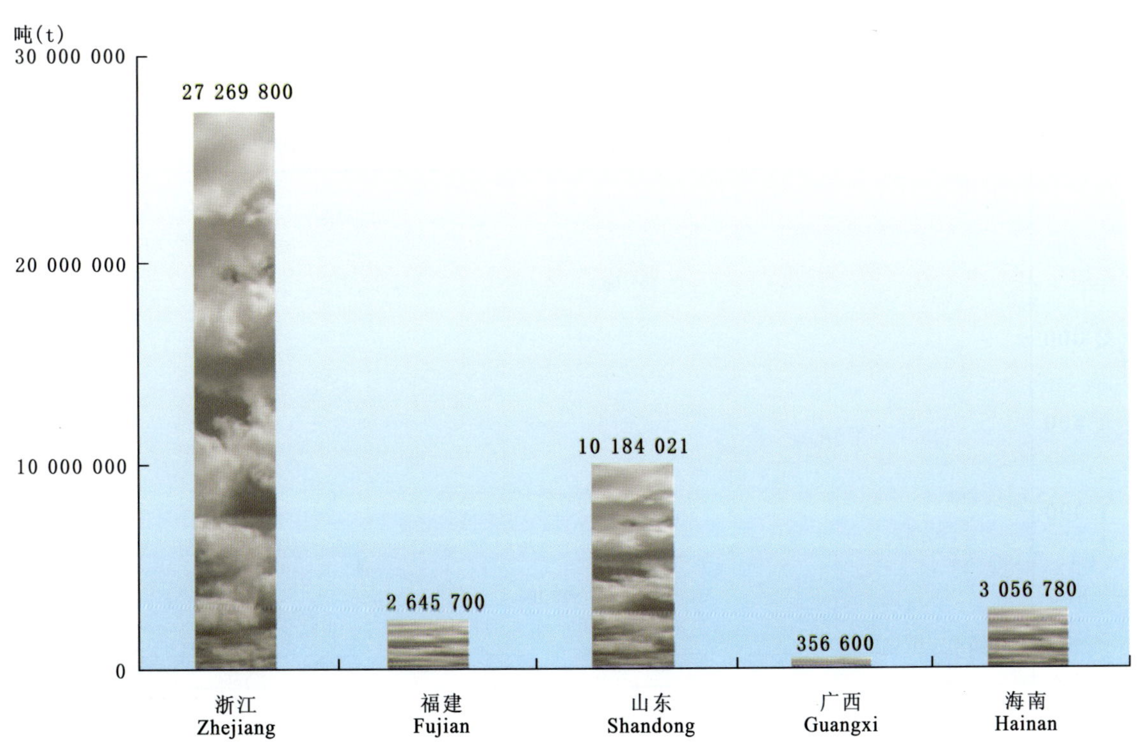

图 9 2012年沿海地区海盐产量

Sea Salt Production by Coastal Regions in 2012

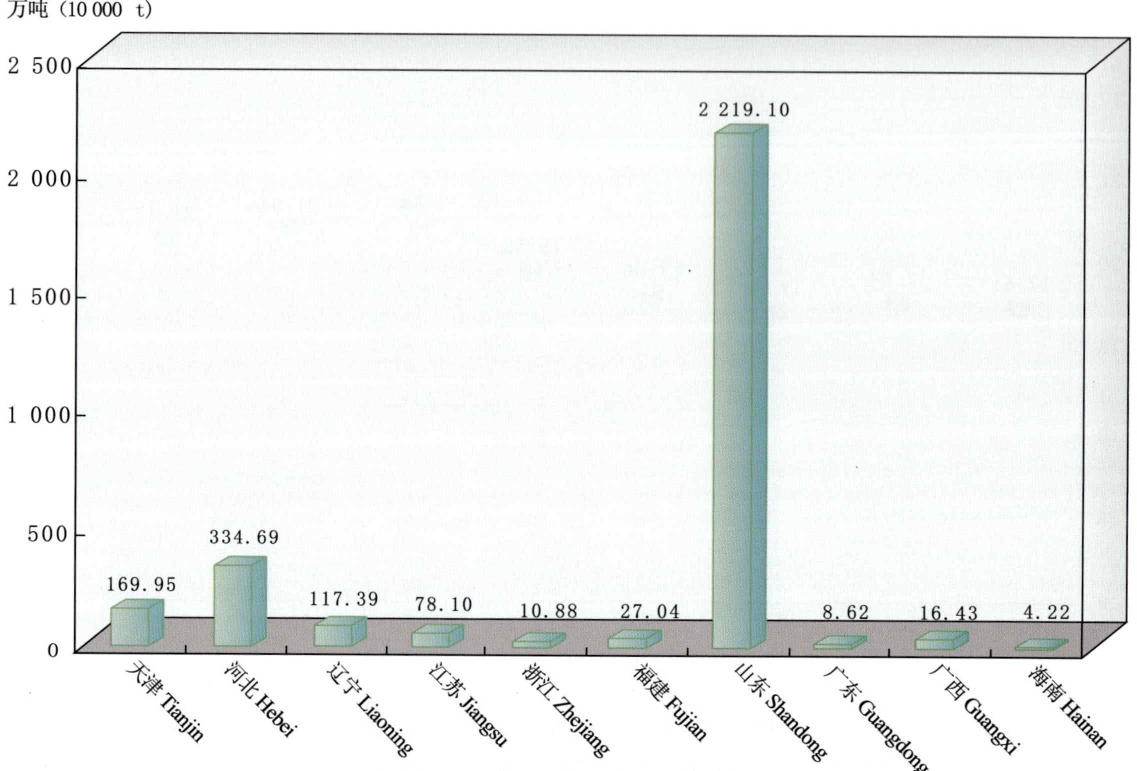

图 10 2012年沿海地区海洋造船完工量

Completed Quantity of Marine Shipbuilding in the Coastal Region in 2012

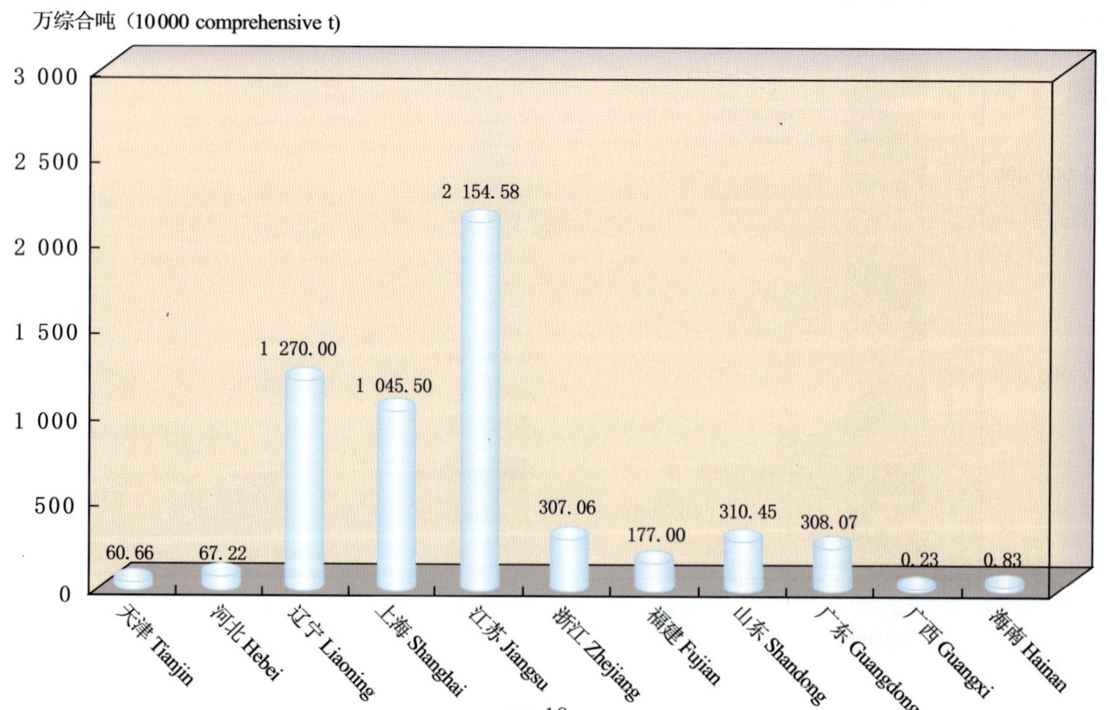

图 11　2012年沿海地区海洋货物周转量
Maritime Goods Turnover Volume by Coastal Regions in 2012

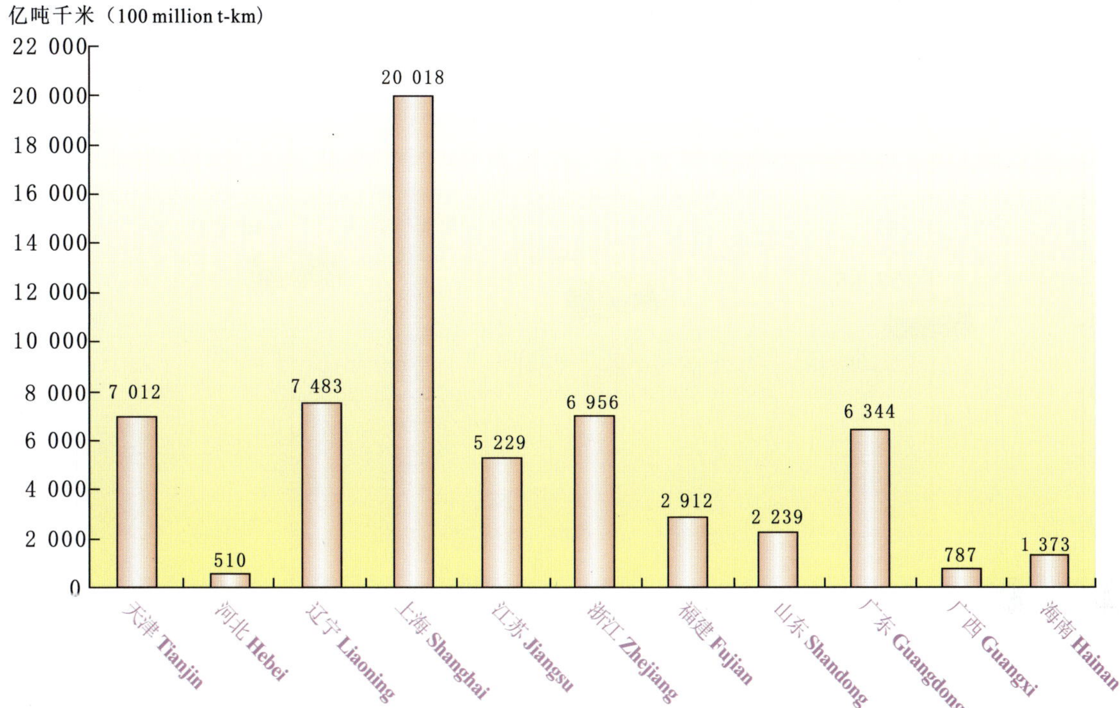

图 12　2012年沿海港口国际标准集装箱吞吐量
International Standardized Containers Handled by Coastal Seaports in 2012

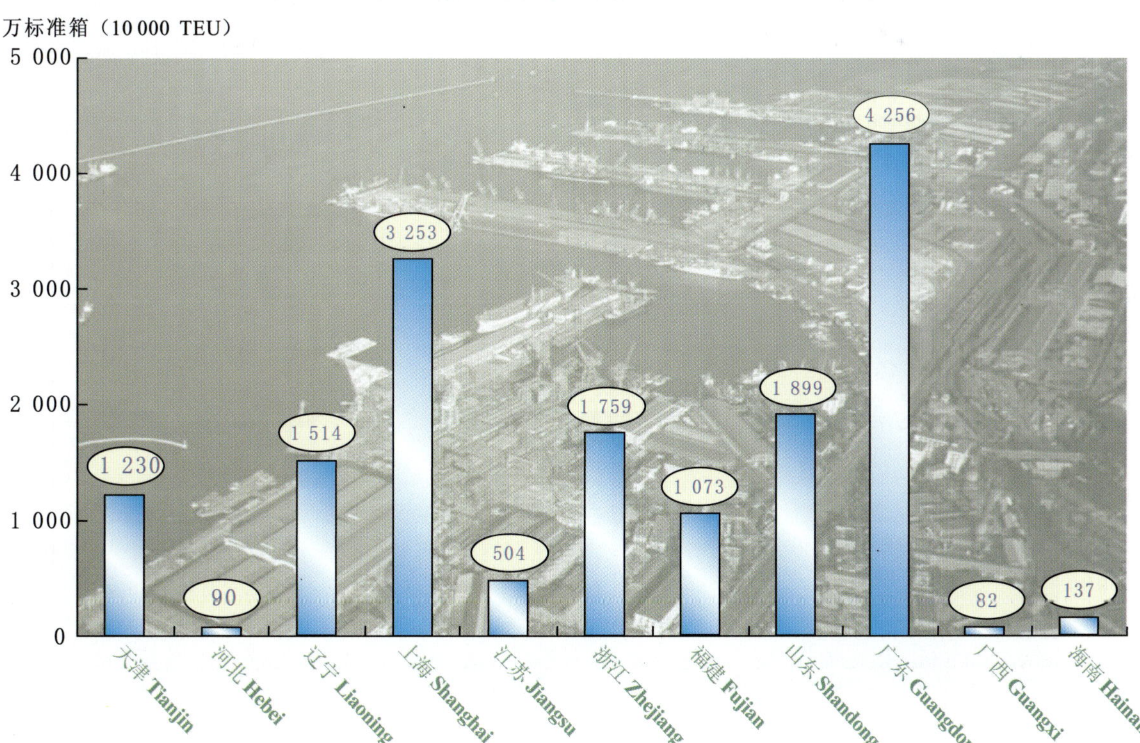

图 13 主要沿海城市国际旅游（外汇）收入
Foreign Exchange Earnings from International Tourism in Major Coastal Cities

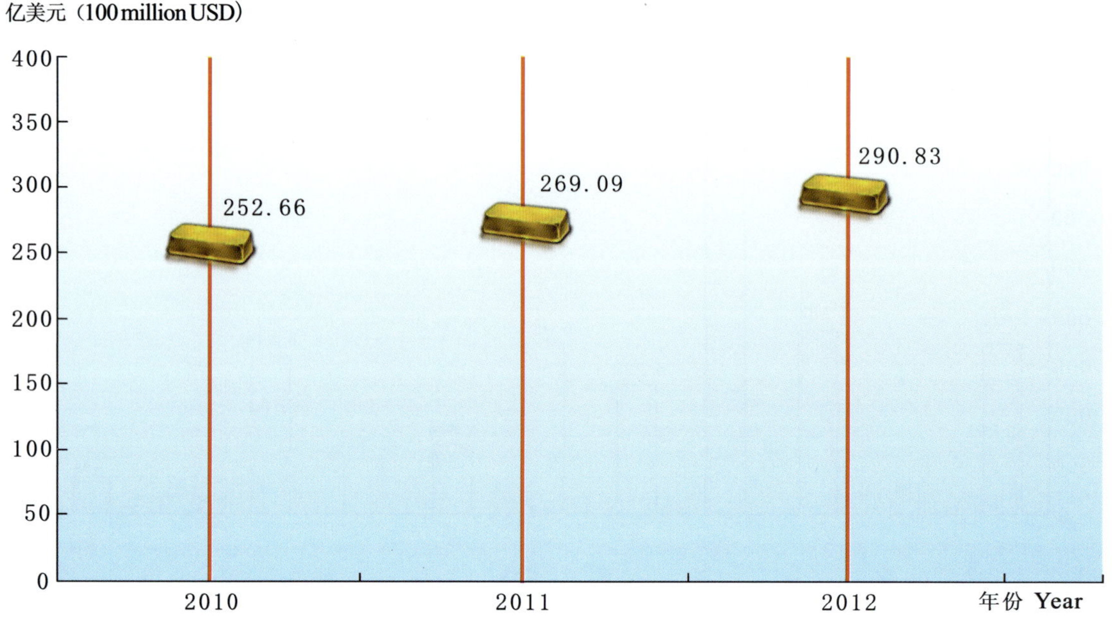

图 14 主要沿海城市接待入境旅游者人数
Number of Oversea Visitor Arrivals in Major Coastal Cities

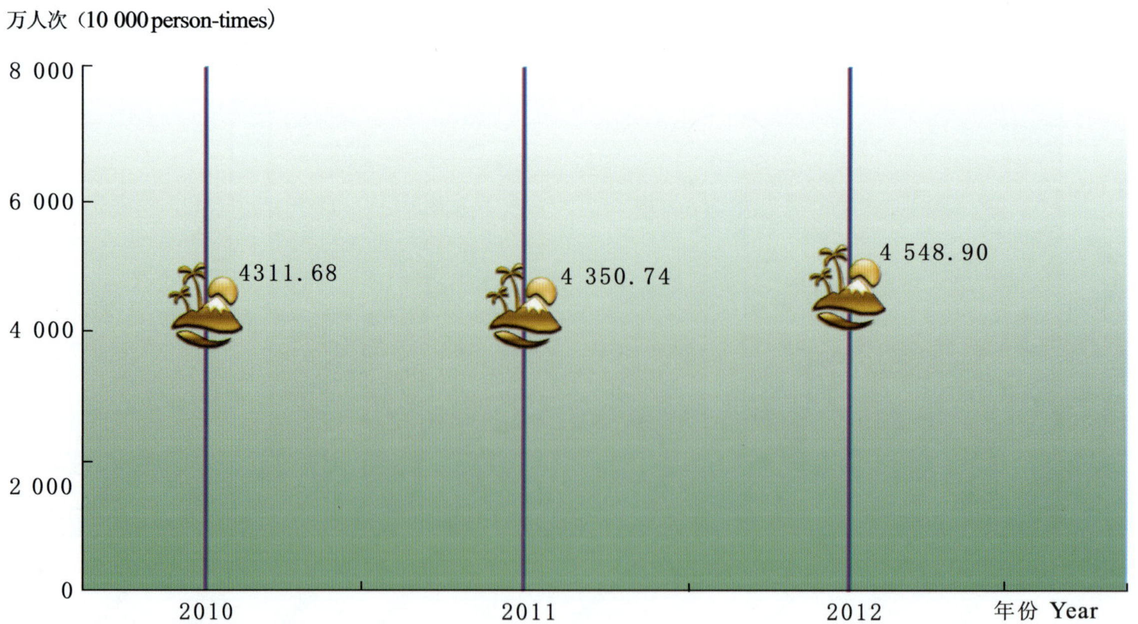

图 15 2012年三大经济区地区生产总值与海洋生产总值

GDP and GOP in the Three Major Economic Zones in 2012

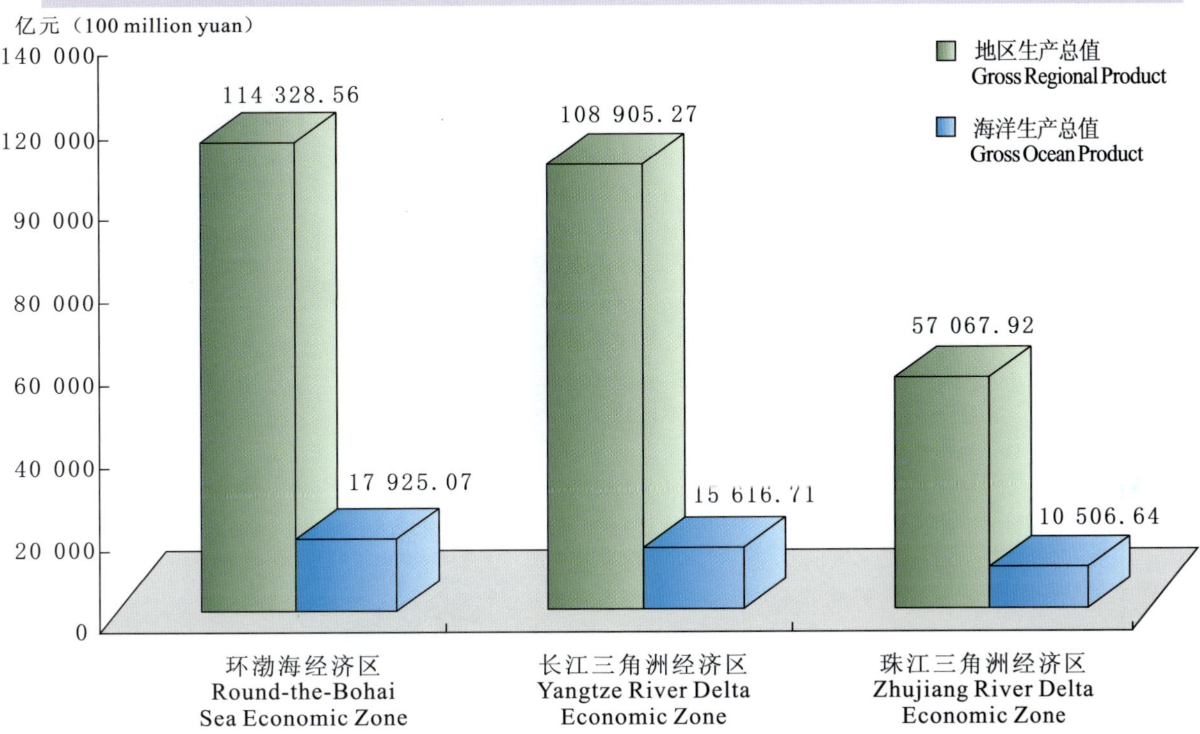

图 16 2012年三大经济区海洋生产总值占全国海洋生产总值比重

Proportion of the GOP of the Three Major Economic Zones in the National GOP in 2012

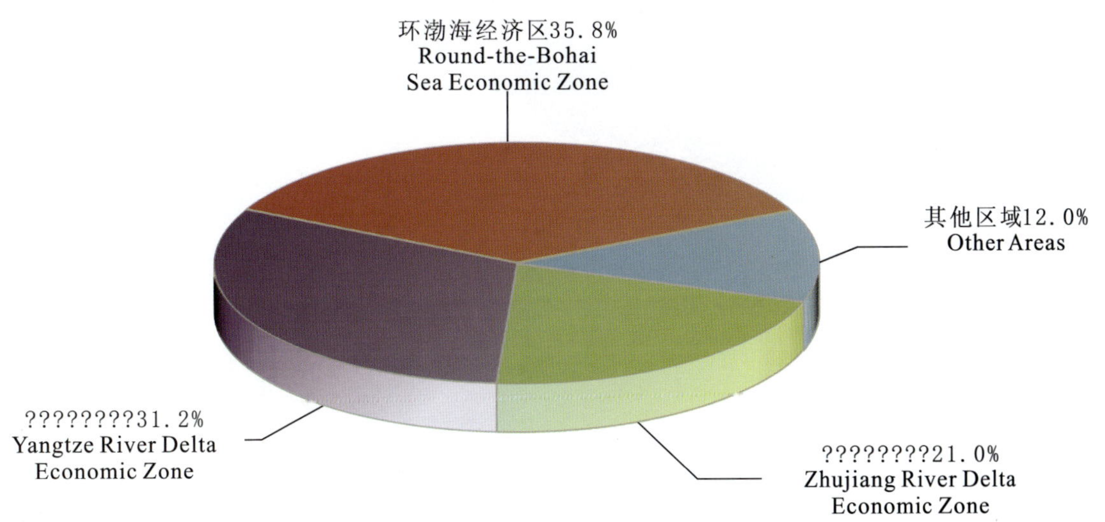

图 17 2012年沿海地区海洋生产总值

Gross Ocean Product by Coastal Regions in 2012

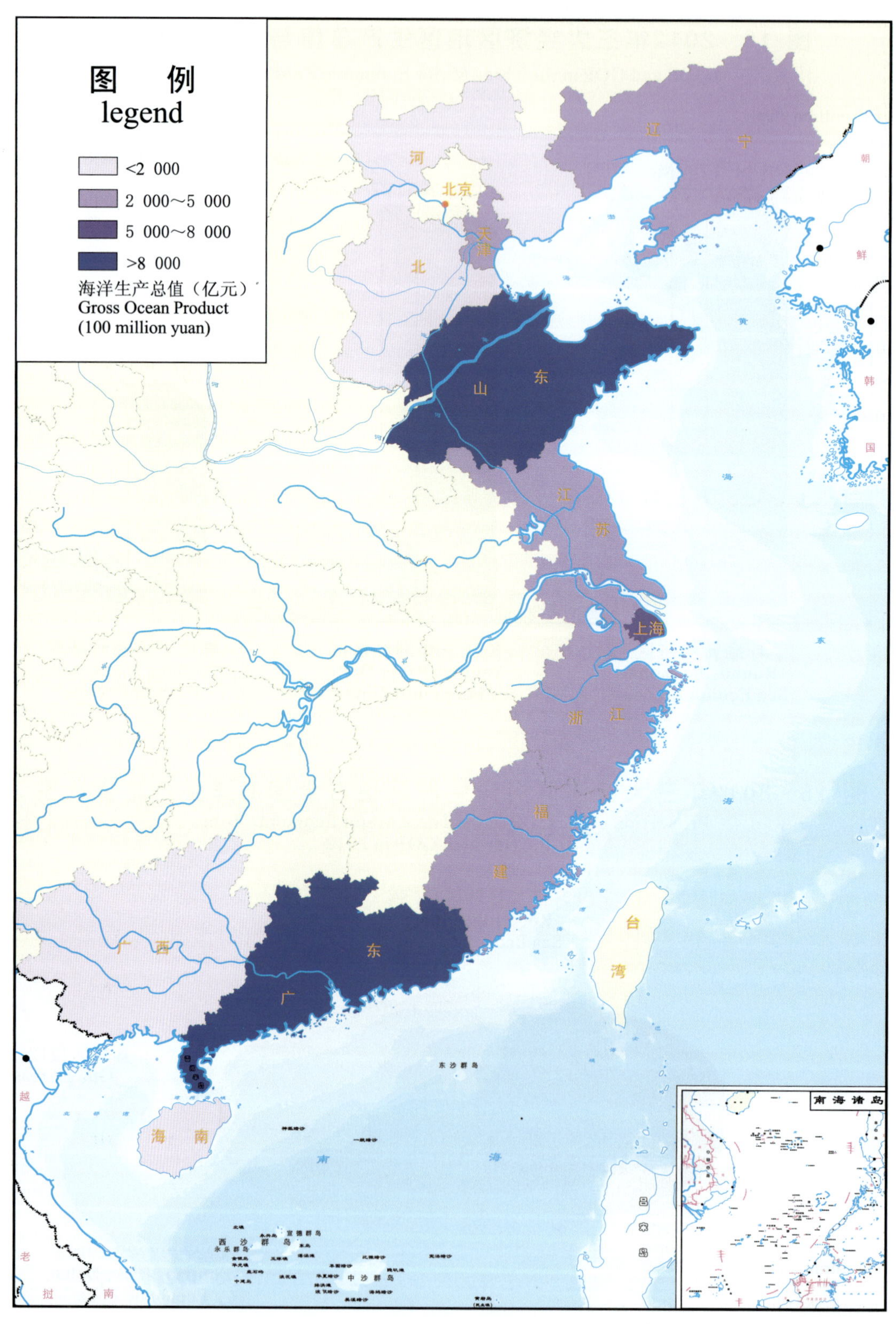

图 18 2012年沿海地区海洋经济贡献

Marine Economic Contributions by Coastal Regions in 2012

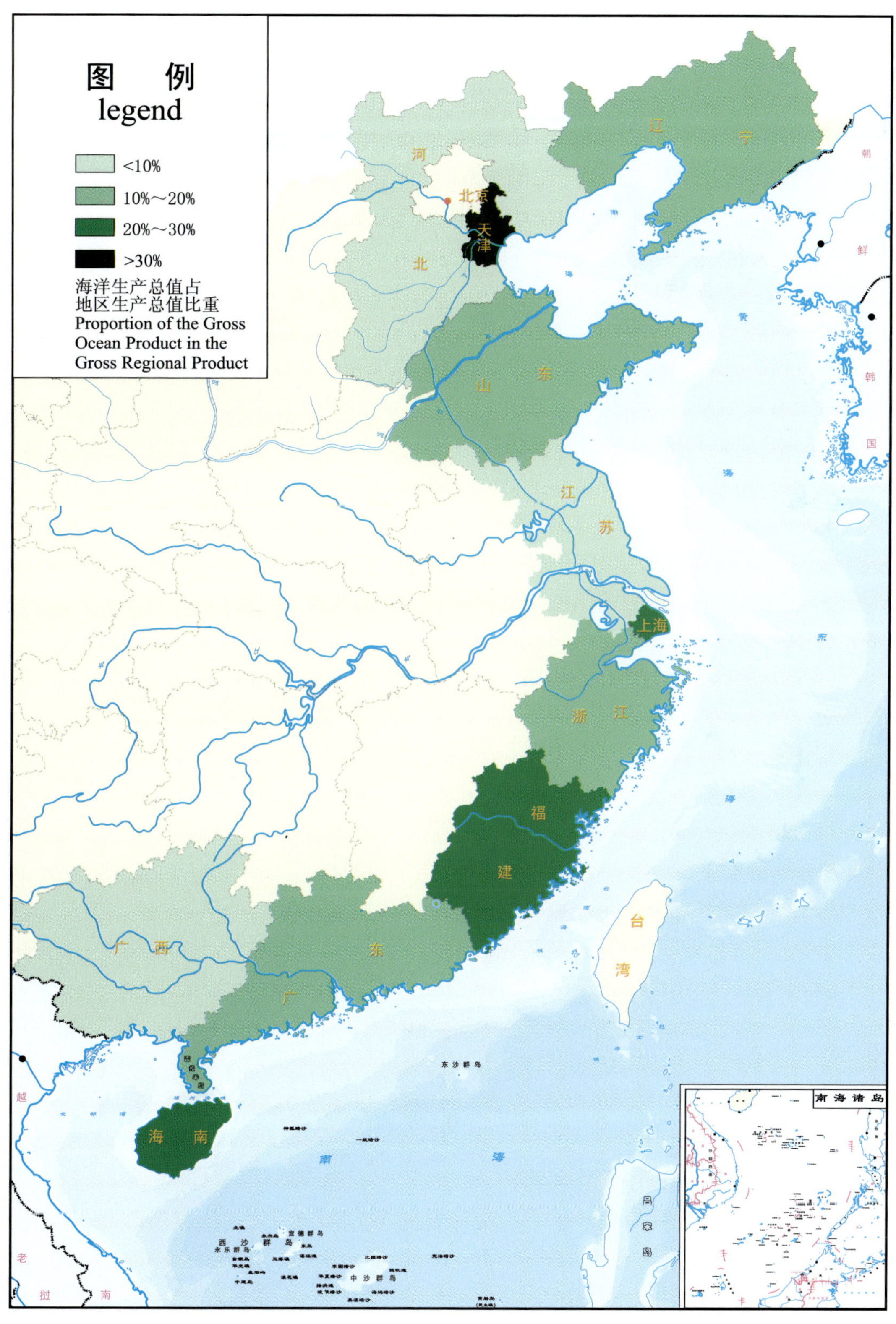

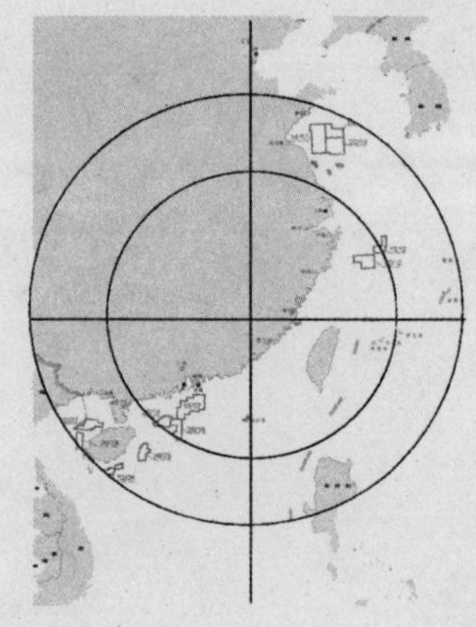

1

综 合 资 料

Integrated Data

1-1 沿海地区行政区划
Administrative Division of Coastal Regions

单位：个 (number)

沿海地区 Coastal Region	沿海城市 Coastal City	沿海地带 Coastal County (District)			
		合计 Total	县 County	县级市 County-level City	区 District
合计 Total	**54**	**236**	**63**	**55**	**118**
天津 Tianjin	1	1			1
河北 Hebei	3	11	6	1	4
辽宁 Liaoning	6	22	4	7	11
上海 Shanghai	1	5	1		4
江苏 Jiangsu	3	15	8	4	3
浙江 Zhejiang	7	35	11	10	14
福建 Fujian	6	34	11	8	15
山东 Shandong	7	35	6	13	16
广东 Guangdong	14	56	10	6	40
广西 Guangxi	3	8	1	1	6
海南 Hainan	3	14	5	5	4

注：沿海地带中未包括广东省的东莞、中山和海南省的三沙市所辖地区。

Note: The coastal zone does not includes the regions under the jurisdiction of Dongguan, Zhongshan, Guangdong Province and Sansha City, Hainan Province.

1-2 沿海行政区划一览表
Table of Administrative Division of Coastal Regions

沿海地区 Coastal Region	地区代码 Zip Code	沿海城市 Coastal City	地区代码 Zip Code	沿海地带 Coastal County (District)	地区代码 Zip Code
天 津 Tianjin	120000			滨海新区Binhai Xinqu	120116
河 北 Hebei	130000	唐山 Tangshan	130200	丰南区Fengnan Qu	130207
				曹妃甸区Caofeidian Qu	130209
				滦南县Luannan Xian	130224
				乐亭县Leting Xian	130225
		秦皇岛 Qinhuangdao	130300	海港区Haigang Qu	130302
				山海关区Shanhaiguan Qu	130303
				北戴河区Beidaihe Qu	130304
				昌黎县Changli Xian	130322
				抚宁县Funing Xian	130323
		沧州 Cangzhou	130900	海兴县Haixing Xian	130924
				黄骅市Huanghua Shi	130983
辽 宁 Liaoning	210000	大连 Dalian	210200	中山区Zhongshan Qu	210202
				西岗区Xigang Qu	210203
				沙河口区Shahekou Qu	210204
				甘井子区Ganjingzi Qu	210211
				旅顺口区Lüshunkou Qu	210212
				金州区Jinzhou Qu	210213
				长海县Changhai Xian	210224
				瓦房店市Wafangdian Shi	210281
				普兰店市Pulandian Shi	210282
				庄河市Zhuanghe Shi	210283
		丹东 Dandong	210600	振兴区Zhenxing Qu	210603
				东港市Donggang Shi	210681
		锦州 Jinzhou	210700	凌海市Linghai Shi	210781
		营口 Yingkou	210800	鲅鱼圈区Bayuquan Qu	210804
				老边区Laobian Qu	210811
				盖州市Gaizhou Shi	210881
		盘锦 Panjin	211100	大洼县Dawa Xian	211121
				盘山县Panshan Xian	211122

1-2 续表1 continued

沿海地区 Coastal Region	地区代码 Zip Code	沿海城市 Coastal City	地区代码 Zip Code	沿海地带 Coastal County (District)	地区代码 Zip Code
		葫芦岛 Huludao	211400	连山区Lianshan Qu	211402
				龙港区Longgang Qu	211403
				绥中县Suizhong Xian	211421
				兴城市Xingcheng Shi	211481
上海 Shanghai	310000			宝山区Baoshan Qu	310113
				浦东新区Pudong Xinqu	310115
				金山区Jinshan Qu	310116
				奉贤区Fengxian Qu	310120
				崇明县Chongming Xian	310230
江苏 Jiangsu	320000	南通 Nantong	320600	通州区Tongzhou Qu	320612
				海安县Hai'an Xian	320621
				如东县Rudong Xian	320623
				启东市Qidong Shi	320681
				海门市Haimen Shi	320684
		连云港 Lianyungang	320700	连云区Lianyun Qu	320703
				新浦区Xinpu Qu	320705
				赣榆县Ganyu Xian	320721
				灌云县Guanyun Xian	320723
				灌南县Guannan Xian	320724
		盐城 Yancheng	320900	响水县Xiangshui Xian	320921
				滨海县Binhai Xian	320922
				射阳县Sheyang Xian	320924
				东台市Dongtai Shi	320981
				大丰市Dafeng Shi	320982
浙江 Zhejiang	330000	杭州 Hangzhou	330100	滨江区Binjiang Qu	330108
				萧山区Xiaoshan Qu	330109
		宁波 Ningbo	330200	海曙区Haishu Qu	330203
				江东区Jiangdong Qu	330204
				江北区Jiangbei Qu	330205
				北仑区Beilun Qu	330206
				镇海区Zhenhai Qu	330211
				鄞州区Yinzhou Qu	330212
				象山县Xiangshan Xian	330225
				宁海县Ninghai Xian	330226
				余姚市Yuyao Shi	330281
				慈溪市Cixi Shi	330282
				奉化市Fenghua Shi	330283

1-2 续表2 continued

沿海地区 Coastal Region	地区代码 Zip Code	沿海城市 Coastal City	地区代码 Zip Code	沿海地带 Coastal County (District)	地区代码 Zip Code
		温州 Wenzhou	330300	龙湾区Longwan Qu	330303
				瓯海区Ouhai Qu	330304
				洞头县Dongtou Xian	330322
				平阳县Pingyang Xian	330326
				苍南县Cangnan Xian	330327
				瑞安市Rui'an Shi	330381
				乐清市Yueqing Shi	330382
		嘉兴 Jiaxing	330400	海盐县Haiyan Xian	330424
				海宁市Haining Shi	330481
				平湖市Pinghu Shi	330482
		绍兴 Shaoxing	330600	绍兴县Shaoxing Xian	330621
				上虞市Shangyu Shi	330682
		舟山 Zhoushan	330900	定海区Dinghai Qu	330902
				普陀区Putuo Qu	330903
				岱山县Daishan Xian	330921
				嵊泗县Shengsi Xian	330922
		台州 Taizhou	331000	椒江区Jiaojiang Qu	331002
				路桥区Luqiao Qu	331004
				玉环县Yuhuan Xian	331021
				三门县Sanmen Xian	331022
				温岭市Wenling Shi	331081
				临海市Linhai Shi	331082
福 建 Fujian	350000	福州 Fuzhou	350100	马尾区Mawei Qu	350105
				连江县Lianjiang Xian	350122
				罗源县Luoyuan Xian	350123
				平潭县Pingtan Xian	350128
				福清市Fuqing Shi	350181
				长乐市Changle Shi	350182
		厦门 Xiamen	350200	思明区Siming Qu	350203
				海沧区Haicang Qu	350205

1-2 续表3 continued

沿海地区 Coastal Region	地区代码 Zip Code	沿海城市 Coastal City	地区代码 Zip Code	沿海地带 Coastal County (District)	地区代码 Zip Code
				湖里区Huli Qu	350206
				集美区Jimei Qu	350211
				同安区Tong'an Qu	350212
				翔安区Xiang'an Qu	350213
		莆田 Putian	350300	城厢区Chengxiang Qu	350302
				涵江区Hanjiang Qu	350303
				荔城区Licheng Qu	350304
				秀屿区Xiuyu Qu	350305
				仙游县Xianyou Xian	350322
		泉州 Quanzhou	350500	丰泽区Fengze Qu	350503
				洛江区Luojiang Qu	350504
				泉港区Quangang Qu	350505
				惠安县Hui'an Xian	350521
				金门县Jinmen Xian	350527
				石狮市Shishi Shi	350581
				晋江市Jinjiang Shi	350582
				南安市Nan'an Shi	350583
		漳州 Zhangzhou	350600	云霄县Yunxiao Xian	350622
				漳浦县Zhangpu Xian	350623
				诏安县Zhao'an Xian	350624
				东山县Dongshan Xian	350626
				龙海市Longhai Shi	350681
		宁德 Ningde	350900	蕉城区Jiaocheng Qu	350902
				霞浦县Xiapu Xian	350921
				福安市Fu'an Shi	350981
				福鼎市Fuding Shi	350982
山 东 Shandong	370000	青岛 Qingdao	370200	市南区Shinan Qu	370202
				市北区Shibei Qu	370203
				黄岛区Huangdao Qu	370211
				崂山区Laoshan Qu	370212
				李沧区Licang Qu	370213
				城阳区Chengyang Qu	370214
				胶州市Jiaozhou Shi	370281
				即墨市Jimo Shi	370282

1-2 续表4 continued

沿海地区 Coastal Region	地区代码 Zip Code	沿海城市 Coastal City	地区代码 Zip Code	沿海地带 Coastal County (District)	地区代码 Zip Code
		东营 Dongying	370500	东营区Dongying Qu	370502
				河口区Hekou Qu	370503
				垦利县Kenli Xian	370521
				利津县Lijin Xian	370522
				广饶县Guangrao Xian	370523
		烟台 Yantai	370600	芝罘区Zhifu Qu	370602
				福山区Fushan Qu	370611
				牟平区Muping Qu	370612
				莱山区Laishan Qu	370613
				长岛县Changdao Xian	370634
				龙口市Longkou Shi	370681
				莱阳市Laiyang Shi	370682
				莱州市Laizhou Shi	370683
				蓬莱市Penglai Shi	370684
				招远市Zhaoyuan Shi	370685
				海阳市Haiyang Shi	370687
		潍坊 Weifang	370700	寒亭区Hanting Qu	370703
				寿光市Shouguang Shi	370783
				昌邑市Changyi Shi	370786
		威海 Weihai	371000	环翠区Huancui Qu	371002
				文登市Wendeng Shi	371081
				荣成市Rongcheng Shi	371082
				乳山市Rushan Shi	371083
		日照 Rizhao	371100	东港区Donggang Qu	371102
				岚山区Lanshan Qu	371103
		滨州 Binzhou	371600	无棣县Wudi Xian	371623
				沾化县Zhanhua Xian	371624
广东 Guangdong	440000	广州 Guangzhou	440100	荔湾区Liwan Qu	440103
				越秀区Yuexiu Qu	440104
				海珠区Haizhu Qu	440105
				天河区Tianhe Qu	440106
				白云区Baiyun Qu	440111
				黄埔区Huangpu Qu	440112
				番禺区Panyu Qu	440113
				南沙区Nansha Qu	440115
				萝岗区Luogang Qu	440116

1-2 续表5 continued

沿海地区 Coastal Region	地区代码 Zip Code	沿海城市 Coastal City	地区代码 Zip Code	沿海地带 Coastal County (District)	地区代码 Zip Code
		深圳 Shenzhen	440300	罗湖区Luohu Qu	440303
				福田区Futian Qu	440304
				南山区Nanshan Qu	440305
				宝安区Bao'an Qu	440306
				龙岗区Longgang Qu	440307
				盐田区Yantian Qu	440308
		珠海 Zhuhai	440400	香洲区Xiangzhou Qu	440402
				斗门区Doumen Qu	440403
				金湾区Jinwan Qu	440404
		汕头 Shantou	440500	龙湖区Longhu Qu	440507
				金平区Jinping Qu	440511
				濠江区Haojiang Qu	440512
				潮阳区Chaoyang Qu	440513
				潮南区Chaonan Qu	440514
				澄海区Chenghai Qu	440583
				南澳县Nan'ao Xian	440523
		江门 Jiangmen	440700	蓬江区Pengjiang Qu	440703
				江海区Jianghai Qu	440704
				新会区Xinhui Qu	440705
				台山市Taishan Shi	440781
				恩平市Enping Shi	440785
		湛江 Zhanjiang	440800	赤坎区Chikan Qu	440802
				霞山区Xiashan Qu	440803
				坡头区Potou Qu	440804
				麻章区Mazhang Qu	440811
				遂溪县Suixi Xian	440823
				徐闻县Xuwen Xian	440825
				廉江市Lianjiang Shi	440881
				雷州市Leizhou Shi	440882
				吴川市Wuchuan Shi	440883
		茂名 Maoming	440900	茂南区Maonan Qu	440902
				茂港区Maogang Qu	440903
				电白县Dianbai Xian	440923
		惠州 Huizhou	441300	惠城区Huicheng Qu	441302
				惠阳区Huiyang Qu	441303
				惠东县Huidong Xian	441323
		汕尾 Shanwei	441500	城 区Chengqu	441502
				海丰县Haifeng Xian	441521
				陆丰市Lufeng Shi	441581

1-2 续表6 continued

沿海地区 Coastal Region	地区代码 Zip Code	沿海城市 Coastal City	地区代码 Zip Code	沿海地带 Coastal County (District)	地区代码 Zip Code
		阳江 Yangjiang	441700	江城区Jiangcheng Qu	441702
				阳西县Yangxi Xian	441721
				阳东县Yangdong Xian	441723
		东莞 Dongguan	441900		
		中山 Zhongshan	442000		
		潮州 Chaozhou	445100	湘桥区Xiangqiao Qu	445102
				饶平县Raoping Xian	445122
		揭阳 Jieyang	445200	榕城区Rongcheng Qu	445202
				揭东区Jiedong Qu	445221
				惠来县Huilai Xian	445224
广 西 Guangxi	450000	北海 Beihai	450500	海城区Haicheng Qu	450502
				银海区Yinhai Qu	450503
				铁山港区Tieshangang Qu	450512
				合浦县Hepu Xian	450521
		防城港 Fangchenggang	450600	港口区Gangkou Qu	450602
				防城区Fangcheng Qu	450603
				东兴市Dongxing Shi	450681
		钦州 Qinzhou	450700	钦南区Qinnan Qu	450702
海 南 Hainan	460000	海口 Haikou	460100	秀英区Xiuying Qu	460105
				龙华区Longhua Qu	460106
				美兰区Meilan Qu	460108
		三亚 Sanya	460200	市辖区Shixia Qu	460201
		三沙 Sansha	460300	西沙群岛Xisha Qundao	460321
				南沙群岛Nansha Qundao	460322
				中沙群岛的岛礁及其海域 Reefs of Zhongsha Qundao and Their Sea Areas	460323
		省直辖县 Counties Directly under the Hainan Province Goverment	469000	琼海市Qionghai Shi	469002
				儋州市Danzhou Shi	469003
				文昌市Wenchang Shi	469005
				万宁市Wanning Shi	469006
				东方市Dongfang Shi	469007
				澄迈县Chengmai Xian	469023
				临高县Lingao Xian	469024
				昌江黎族自治县Changjiang Lizu Zizhixian	469026
				乐东黎族自治县 Ledong Lizu Zizhixian	469027
				陵水黎族自治县 Lingshui Lizu Zizhixian	469028

1-3 海洋自然地理
Marine Physical Geography

指　　标		Item	指标值 Data
海洋平均深度	（米）	Average Depth of Sea (m)	961
海洋最大深度	（米）	Maximum Depth of Sea (m)	5 559
岸线总长度	（千米）	Total Length of Coastline (km)	32 000
大陆岸线长度		Length of Continental Coastline	18 000
岛屿岸线长度		Length of Insular Coastline	14 000
＞$500m^2$岛屿数	（个）	Number of Islands ＞500 m^2 each (unit)	7 300
岛屿面积	（万平方千米）	Area of Islands (10 000 km^2)	8
已利用的无居民海岛	（个）	Uninhabited Islands Which Have Been Utilized (unit)	1 900
特殊用途海岛		Islands Used for Special Purposes	1 020
公共服务用岛		Islands Used for Public Service	365
旅游娱乐用岛		Islands Used for Tourism and Recreation	73
农林牧渔业用岛		Islands Used for Agriculture, Forestry, Animal Husbandry and Fishery	340
工业、仓储、交通运输用岛		Islands Used for Industry, Storage, Communications and Transport	49

注：海岛数据来源于《全国海岛保护规划》。

Note: Data on islands are derived from the *National Plan for Island Protection*.

1-4　海区海洋石油储量
Offshore Oil Reserves in the Sea Area

自然海区名称 Natural Sea Area	海洋石油（万吨） Offshore Oil (10 000 t)	
	累计探明技术可采储量 Proven Technically Recoverable Reserves in the Aggregate	剩余技术可采储量 Surplus Technically Recoverable Reserves
合 计　Total	**94 267.2**	**48 005.7**
渤 海　Bohai Sea	56 246.9	35 088.0
黄 海　Huanghai Sea		
东 海　Donghai Sea	1 432.2	1 015.2
南 海　Nanhai Sea	36 588.1	11 902.5

注：数据来源于《2012年全国矿产资源储量通报》。

Note: The data come from the *Journal on the National Mineral Resources Reserves in 2012*.

1-5 沿海地区水资源情况
Water Resources by Coastal Regions

地 区 Region	水资源总量（亿立方米） Total Amount of Water Resources (100 million m^3)				人均水资源量（立方米/人） Per Capita Water Resources (m^3/person)
		地表水资源量 Surface Water Resources	地下水资源量 Groundwater Resources	地表水与地下水资源重复量 Duplicate Measurement of Surface Water and Groundwater	
全国总计 National Total	**29 526.9**	**28 371.4**	**8 416.1**	**7 260.6**	**2 186.1**
天 津 Tianjin	32.9	26.5	7.6	1.2	238.0
河 北 Hebei	235.5	117.8	164.8	47.1	324.2
辽 宁 Liaoning	547.3	492.4	147.4	92.5	1 247.8
上 海 Shanghai	33.9	27.4	9.7	3.2	143.4
江 苏 Jiangsu	373.3	279.1	110.2	16.0	472.0
浙 江 Zhejiang	1 444.8	1 427.1	273.5	255.8	2 641.3
福 建 Fujian	1 511.4	1 510.1	349.3	347.9	4 047.8
山 东 Shandong	274.3	182.2	164.2	72.1	283.9
广 东 Guangdong	2 026.5	2 017.5	485.8	476.7	1 921.0
广 西 Guangxi	2 087.4	2 086.4	587.3	586.3	4 476.0
海 南 Hainan	364.3	360.2	92.6	88.5	4 130.8

注： 数据来源于《2013中国统计年鉴》。

Note: The data come from the *China Statistical Yearbook 2013*.

1-6 沿海地区湿地面积
Area of Wetlands by Coastal Regions

地 区 Region	湿地面积 （千公顷） Area of Wetlands (1 000 hm²)	近岸及海岸 Inshore Areas and Seashores	湿地面积 占国土面积 比重 （%） Proportion of Wetlands in the Total Area of Territory (%)
全国总计 National Total	**38 485.5**	**5 941.7**	**4.01**
天 津 Tianjin	171.8	58.1	14.95
河 北 Hebei	1 081.9	278.8	5.82
辽 宁 Liaoning	1 219.6	738.1	8.37
上 海 Shanghai	319.7	305.4	53.68
江 苏 Jiangsu	1 674.7	843.5	16.32
浙 江 Zhejiang	802.2	574.3	7.88
福 建 Fujian	443.0	370.6	3.65
山 东 Shandong	1 784.1	1 210.9	11.72
广 东 Guangdong	1 398.1	1 017.8	7.86
广 西 Guangxi	656.1	348.4	2.76
海 南 Hainan	311.5	190.0	9.13

注： 数据来源于《2013中国统计年鉴》。

Note: The data come from the *China Statistical Yearbook 2013*.

1-7 红树林各地类面积
Site Classification and Area of Sharpleaf Mangrove (*Rhizophora Apiculata*)

单位：公顷 (hm^2)

地 区 Region	红树林各地类总面积 Total Site Area of Sharpleaf Mangrove	现有面积 Established	未成林面积 Unestablished	宜林地面积 Suitable for Planting
全国总计 National Total	**82 757.2**	**22 024.9**	**1 884.1**	**58 848.2**
浙 江 Zhejiang	5 452.3	20.6	236.1	5 195.6
福 建 Fujian	13 410.1	615.1	286.4	12 508.6
广 东 Guangdong	32 325.9	9 084.0	981.3	22 260.6
广 西 Guangxi	18 029.2	8 374.9	380.3	9 274.0
海 南 Hainan	13 539.7	3 930.3		9 609.4

注：数据来源于《2013中国统计年鉴》。

Note: The data come from the *China Statistical Yearbook 2013*.

1-8 主要沿海城市气候基本情况
Climate of Major Coastal Cities

城 市 City	年平均气温 （摄氏度） Annual Average Temperature(℃)	年平均相对湿度 （%） Annual Average Relative Humidity (%)	全年降水量 （毫米） Annual Precipitation (mm)	全年日照时数 （小时） Annual Sunshine Hours (h)
天 津 Tianjin	12.5	57	755.3	2 174.4
上 海 Shanghai	16.9	70	1 103.7	1 676.7
杭 州 Hangzhou	17.1	71	1 728.8	1 520.5
福 州 Fuzhou	20.2	75	1 913.4	1 291.3
广 州 Guangzhou	21.7	82	1 813.9	1 471.2
海 口 Haikou	24.6	82	2 094.3	1 766.4

注：数据来源于《2013中国统计年鉴》。

Note: The data come from the *China Statistical Yearbook 2013*.

主要统计指标解释

1. 沿海地区 即广义的沿海地区，是指有海岸线（大陆岸线和岛屿岸线）的地区，按行政区划分为沿海省、自治区、直辖市。

2. 沿海城市 是指有海岸线的直辖市和地级市（包括其下属的全部区、县和县级市）。

3. 沿海地带 即狭义的沿海地区，是指有海岸线的县、县级市、区（包括直辖市和地级市的区）。

4. 海洋 是海和洋的统称。洋为地球表面上相连接的广大咸水水体的主体部分。海为地球表面相连接的广大咸水水体被陆地、岛礁、半岛包围或分隔的边缘部分。

5. 水资源总量 指评价区内降水形成的地表和地下产水总量，即地表产流量与降水入渗补给地下水量之和，不包括过境水量。

6. 地表水资源量 指评价区内河流、湖泊、冰川等地表水体中可以逐年更新的动态水量，即当地天然河川径流量。

7. 地下水资源量 指评价区内降水和地表水对饱水岩土层的补给量，包括降水入渗补给量和河道、湖库、渠系、渠灌田间等地表水体的入渗补给量。

8. 地表水与地下水资源重复量 指地表水和地下水相互转化的部分，即天然河川径流量中的地下水排泄量和地下水补给量中来源于地表水的入渗补给量。

9. 湿地 指天然或人工、长久或暂时性的沼泽地、泥炭地或水域地带，包括静止或流动、淡水、半咸水、咸水体，低潮时水深不超过6米的水域以及海岸地带地区的珊瑚滩和海草床、滩涂、红树林、河口、河流、淡水沼泽、沼泽森林、湖泊、盐沼及盐湖。

10. 红树林 指生长在热带、亚热带低能海岸潮间带上部，受周期性潮水浸淹，以红树植物为主体的常绿灌木或乔木组成的潮滩湿地木本生物群落。

11. 气温 指空气的温度，我国一般以摄氏度(℃)为单位表示。气象观测的温度表是放在离地面约1.5米处通风良好的百叶箱里测量的，因此，通常说的气温指的是离地面1.5米处百叶箱中的温度。其统计计算方法为：

月平均气温是将全月各日的平均气温相加，除以该月的天数而得。

年平均气温是将12个月的月平均气温累加后除以12而得。

12. 相对湿度 指空气中实际所含水蒸气密度和同温度下饱和水蒸气密度的百分比值。其统计方法与气温相同。

13. 降水量 指从天空降落到地面的液态或固态(经融化后)水，未经蒸发、渗透、流失而在地面上积聚的深度。其统计计算方法为：

月降水量是将全月各日的降水量累加而得。

年降水量是将12个月的月降水量累加而得。

14. 日照时数 指太阳实际照射地面的时间。其统计方法与降水量相同。

Explanatory Notes on Main Statistical Indicators

1. Coastal Region, i.e., the coastal region in a broad sense, refers to the regions with coastlines (continental and island coastlines), which are divided into the coastal provinces, autonomous regions and municipalities directly under the Central Government according to the administrative zoning.

2. Coastal City refers to the municipalities directly under the Central Government and the prefecture-level cities (including all the districts, counties and county-level cities under them).

3. Coastal Zone, i.e., the coastal region in a narrow sense, refers to the counties, county-level cities and districts with coastlines (including the districts under the municipalities directly under the Central Government and the prefecture-level districts).

4. Ocean is the general name for sea and ocean. Ocean refers to the main body of large salt water connected with the earth surface. Sea refers to the edge areas of the salt water on the earth surface that are compartmentalized or surrounded by land, island, reef or peninsula.

5. Total Water Resources refers to total volume of water resources measured as run-off for surface water from rainfall and recharge for groundwater in a given area, excluding transit water.

6. Surface Water Resources refers to total renewable resources which exist in rivers, lakes, glaciers and other collectors from rainfall and are measured as run-off of rivers.

7. Groundwater Resources refers to replenishment of aquifers with rainfall and surface water.

8. Duplicated Measurement between Surface Water and Groundwater refers to the exchange between surface water and groundwater, i.e. run-off of rivers includes some depletion into groundwater while groundwater includes some replenishment from surface water.

9. Wetlands refer to marshland and peat bog, whether natural or man-made, permanent or temporary; water covered areas, whether stagnant or flowing, with fresh or brackish-fresh or salty water that is less than 6 meters deep at low tide; as well as coral beach, weed beach, mud beach, mangrove, river outlet, rivers, fresh-water marshland, marshland forests, lakes, salty bog and salt lakes along the coastal areas.

10. Mangrove refers to evergreen woody plants or plant communities in tropical or sub-tropical zones which live between the sea and the land in areas which are inundated by tides.

11. Temperature refers to the air temperature. China uses centigrade as the unit. The thermometry used for weather observation is put in a breezy shutter, which is 1.5 meters high from the ground. Therefore, the commonly used temperature refers to the temperature in the breezy shutter 1.5 meters away from the ground. The calculation method is as follows:

Monthly average temperature is the summation of average daily temperature of one month divided by the actual days of that particular month.

Annual average temperature is the summation of monthly averages of a year divided by 12 months.

12. Relative Humidity refers to the ratio of actual water vapour pressure to the saturated water vapour density under the current temperature. The calculation method is the same as that of temperature.

13. Volume of Precipitation refers to the deepness of liquid state or solid state (thawed) water falling from the sky to the ground that has not evaporated, infiltrated or run off. The calculation method is as follows:

Monthly precipitation is the summation of daily precipitation of a month.

Annual precipitation is the summation of 12 months precipitation of a year.

14. Sunshine Hours refer to the actual hours of sun irradiating the earth. The calculation method is the same as that of the precipitation.

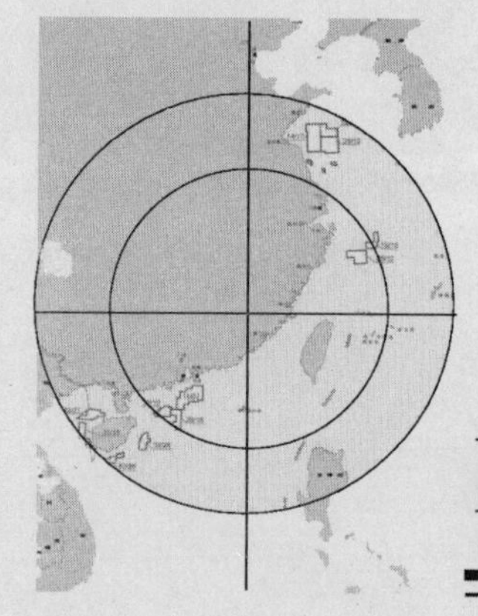

2

海洋经济核算

Marine Economic Accounting

2-1 全国海洋生产总值
National Gross Ocean Product

年 份 Year	海洋生产总值（亿元） Gross Ocean Product (100 million yuan)				海洋生产总值占国内生产总值比重（%） Proportion of the Gross Ocean Product in GDP (%)	海洋生产总值增长速度（%） Growth Rate of the Gross Ocean Product (%)
		第一产业 Primary Industry	第二产业 Secondary Industry	第三产业 Tertiary Industry		
2001	9 518.4	646.3	4 152.1	4 720.1	8.68	
2002	11 270.5	730.0	4 866.2	5 674.3	9.37	19.8
2003	11 952.3	766.2	5 367.6	5 818.5	8.80	4.2
2004	14 662.0	851.0	6 662.8	7 148.2	9.17	16.9
2005	17 655.6	1 008.9	8 046.9	8 599.8	9.55	16.3
2006	21 592.4	1 228.8	10 217.8	10 145.7	9.98	18.0
2007	25 618.7	1 395.4	12 011.0	12 212.3	9.64	14.8
2008	29 718.0	1 694.3	13 735.3	14 288.4	9.46	9.9
2009	32 277.6	1 857.7	14 980.3	15 439.5	9.47	9.2
2010	39 572.7	2 008.0	18 935.0	18 629.8	9.86	14.7
2011	45 496.0	2 381.9	21 685.6	21 428.5	9.62	9.9
2012	50 045.2	2 670.6	23 469.8	23 904.8	9.64	8.1

2-2 全国海洋生产总值构成
Composition of National Gross Ocean Product

单位：% (%)

年 份 Year	第一产业 Primary Industry	第二产业 Secondary Industry	第三产业 Tertiary Industry
2001	6.8	43.6	49.6
2002	6.5	43.2	50.3
2003	6.4	44.9	48.7
2004	5.8	45.4	48.8
2005	5.7	45.6	48.7
2006	5.7	47.3	47.0
2007	5.4	46.9	47.7
2008	5.7	46.2	48.1
2009	5.8	46.4	47.8
2010	5.1	47.8	47.1
2011	5.2	47.7	47.1
2012	5.3	46.9	47.8

2-3　海洋及相关产业增加值
Added Values of Marine and Related Industries

单位：亿元 (100 million yuan)

年份 Year	合计 Total	海洋产业 Marine Industry	主要海洋产业 Major Marine Industry	海洋科研教育管理服务业 Industries of Marine Scientific Research, Education, Management and Service	海洋相关产业 Ocean-related Industries
2001	9 518.4	5 733.6	3 856.6	1 877.0	3 784.8
2002	11 270.5	6 787.3	4 696.8	2 090.5	4 483.2
2003	11 952.3	7 137.7	4 754.4	2 383.3	4 814.6
2004	14 662.0	8 710.1	5 827.7	2 882.5	5 951.9
2005	17 655.6	10 539.0	7 188.0	3 350.9	7 116.6
2006	21 592.4	12 696.7	8 790.4	3 906.4	8 895.6
2007	25 618.7	15 070.6	10 478.3	4 592.3	10 548.0
2008	29 718.0	17 591.2	12 176.0	5 415.2	12 126.8
2009	32 277.6	18 822.0	12 843.6	5 978.4	13 455.6
2010	39 572.7	22 831.0	16 187.8	6 643.1	16 741.7
2011	45 496.0	26 422.0	18 865.2	7 556.8	19 074.1
2012	50 045.2	29 264.4	20 829.9	8 434.6	20 780.8

2-4　海洋及相关产业增加值构成
Composition of the Added Values of Marine and Related Industries

单位：% (%)

年份 Year	合计 Total	海洋产业 Marine Industry	主要海洋产业 Major Marine Industry	海洋科研教育管理服务业 Industries of Marine Scientific Research, Education, Management and Service	海洋相关产业 Ocean-related Industries
2001	100.0	60.2	40.5	19.7	39.8
2002	100.0	60.2	41.7	18.5	39.8
2003	100.0	59.7	39.8	19.9	40.3
2004	100.0	59.4	39.7	19.7	40.6
2005	100.0	59.7	40.7	19.0	40.3
2006	100.0	58.8	40.7	18.1	41.2
2007	100.0	58.8	40.9	17.9	41.2
2008	100.0	59.2	41.0	18.2	40.8
2009	100.0	58.3	39.8	18.5	41.7
2010	100.0	57.7	40.9	16.8	42.3
2011	100.0	58.1	41.5	16.6	41.9
2012	100.0	58.5	41.6	16.9	41.5

2-5 全国主要海洋产业增加值
Added Values of National Major Marine Industries

主要海洋产业 Major Marine Industry	增加值 （亿元） Added Value (100 million yuan)	比上年增长（%） (按可比价计算) Percentage of Increase Over Last Year (%) (at comparable price)
合计 **Total**	**20 829.9**	**7.0**
海洋渔业 Marine Fishery Industry	3 560.5	4.6
海洋油气业 Offshore Oil and Natural Gas Industry	1 718.7	0.2
海洋矿业 Marine Mining Industry	45.1	- 13.5
海洋盐业 Sea Salt Industry	60.1	- 24.3
海洋船舶工业 Marine Shipbuilding Industry	1 291.3	- 4.0
海洋化工业 Marine Chemical Industry	843.0	26.2
海洋生物医药业 Marine Biomedicine Industry	184.7	22.3
海洋工程建筑业 Marine Engineering Architecture Industry	1 353.8	22.6
海洋电力业 Marine Electric Power Industry	77.3	25.9
海水利用业 Marine Seawater Utilization Industry	11.1	3.8
海洋交通运输业 Maritime Communications and Transportation Industry	4 752.6	5.2
滨海旅游业 Coastal Tourism	6 931.8	8.9

2-6 海洋渔业增加值
Added Value of Marine Fishery Industry

单位：亿元 (100 million yuan)

年 份 Year	增加值 Added Value
2001	966.0
2002	1 091.2
2003	1 145.0
2004	1 271.2
2005	1 507.6
2006	1 672.0
2007	1 906.0
2008	2 228.6
2009	2 440.8
2010	2 851.6
2011	3 202.9
2012	3 560.5

2-7 海洋油气业增加值
Added Value of Offshore Oil and Gas Industry

单位：亿元 (100 million yuan)

年 份 Year	增加值 Added Value
2001	176.8
2002	181.8
2003	257.0
2004	345.1
2005	528.2
2006	668.9
2007	666.9
2008	1 020.5
2009	614.1
2010	1 302.2
2011	1 719.7
2012	1 718.7

2-8　海洋矿业增加值
Added Value of Marine Mining Industry

单位：亿元　　(100 million yuan)

年　份 Year	增加值 Added Value
2001	1.0
2002	1.9
2003	3.1
2004	7.9
2005	8.3
2006	13.4
2007	16.3
2008	35.2
2009	41.6
2010	45.2
2011	53.3
2012	45.1

注：自2008年起部分地区统计矿种增加。

Note: The data from 2008 include added kinds of minerals in some regions.

2-9　海洋盐业增加值
Added Value of Marine Salt Industry

单位：亿元　　(100 million yuan)

年　份 Year	增加值 Added Value
2001	32.6
2002	34.2
2003	28.4
2004	39.0
2005	39.1
2006	37.1
2007	39.9
2008	43.6
2009	43.6
2010	65.5
2011	76.8
2012	60.1

2-10 海洋船舶工业增加值
Added Value of Marine Shipbuilding Industry

单位：亿元 (100 million yuan)

年 份 Year	增加值 Added Value
2001	109.3
2002	117.4
2003	152.8
2004	204.1
2005	275.5
2006	339.5
2007	524.9
2008	742.6
2009	986.5
2010	1 215.6
2011	1 352.0
2012	1 291.3

2-11 海洋化工业增加值
Added Value of Marine Chemical Industry

单位：亿元 (100 million yuan)

年 份 Year	增加值 Added Value
2001	64.7
2002	77.1
2003	96.3
2004	151.5
2005	153.3
2006	440.4
2007	506.6
2008	416.8
2009	465.3
2010	613.8
2011	695.9
2012	843.0

注：自2006年起部分地区统计产品品种增加。
Note: The data from 2006 include added kinds of statistical products in some regions.

2-12 海洋生物医药业增加值
Added Value of Marine Biomedicine Industry

单位：亿元 (100 million yuan)

年 份 Year	增加值 Added Value
2001	5.7
2002	13.2
2003	16.5
2004	19.0
2005	28.6
2006	34.8
2007	45.4
2008	56.6
2009	52.1
2010	83.8
2011	150.8
2012	184.7

2-13 海洋工程建筑业增加值
Added Value of Marine Engineering Architecture

单位：亿元 (100 million yuan)

年 份 Year	增加值 Added Value
2001	109.2
2002	145.4
2003	192.6
2004	231.8
2005	257.2
2006	423.7
2007	499.7
2008	347.8
2009	672.3
2010	874.2
2011	1 086.8
2012	1 353.8

2-14　海洋电力业增加值
Added Value of Marine Electric Power Industry

单位：亿元　　(100 million yuan)

年　份 Year	增加值 Added Value
2001	1.8
2002	2.2
2003	2.8
2004	3.1
2005	3.5
2006	4.4
2007	5.1
2008	11.3
2009	20.8
2010	38.1
2011	59.2
2012	77.3

2-15　海水利用业增加值
Added Value of Seawater Utilization Industry

单位：亿元　　(100 million yuan)

年　份 Year	增加值 Added Value
2001	1.1
2002	1.3
2003	1.7
2004	2.4
2005	3.0
2006	5.2
2007	6.2
2008	7.4
2009	7.8
2010	8.9
2011	10.4
2012	11.1

2-16　海洋交通运输业增加值
Added Value of Marine Communications and Transportation Industry

单位：亿元　　(100 million yuan)

年　份 Year	增加值 Added Value
2001	1 316.4
2002	1 507.4
2003	1 752.5
2004	2 030.7
2005	2 373.3
2006	2 531.4
2007	3 035.6
2008	3 499.3
2009	3 146.6
2010	3 785.8
2011	4 217.5
2012	4 752.6

2-17　滨海旅游业增加值
Added Value of Coastal Tourism

单位：亿元　　(100 million yuan)

年　份 Year	增加值 Added Value
2001	1 072.0
2002	1 523.7
2003	1 105.8
2004	1 522.0
2005	2 010.6
2006	2 619.6
2007	3 225.8
2008	3 766.4
2009	4 352.3
2010	5 303.1
2011	6 239.9
2012	6 931.8

2-18 沿海地区海洋生产总值
Gross Ocean Product by Coastal Regions

地 区 Region	海洋生产总值（亿元） Gross Ocean Product (100 million yuan)				海洋生产总值占沿海地区生产总值比重（%） Proportion of the Gross Ocean Product in the Gross Regional Product（%）
		第一产业 Primary Industry	第二产业 Secondary Industry	第三产业 Tertiary Industry	
合 计 **Total**	**50 045.2**	**2 670.6**	**23 469.8**	**23 904.8**	**15.7**
天 津 Tianjin	3 939.2	7.9	2 626.0	1 305.3	30.6
河 北 Hebei	1 622.0	70.9	876.3	674.7	6.1
辽 宁 Liaoning	3 391.7	447.0	1 339.7	1 605.1	13.7
上 海 Shanghai	5 946.3	4.2	2 248.2	3 693.9	29.5
江 苏 Jiangsu	4 722.9	220.4	2 439.2	2 063.4	8.7
浙 江 Zhejiang	4 947.5	369.7	2 180.4	2 397.4	14.3
福 建 Fujian	4 482.8	416.3	1 815.9	2 250.7	22.8
山 东 Shandong	8 972.1	648.7	4 362.8	3 960.6	17.9
广 东 Guangdong	10 506.6	180.1	5 134.9	5 191.7	18.4
广 西 Guangxi	761.0	142.7	301.8	316.5	5.8
海 南 Hainan	752.9	162.7	144.6	445.6	26.4

2-19 沿海地区海洋生产总值构成
Composition of Gross Ocean Product by Coastal Regions

单位：% (%)

地 区 Region	海洋生产总值 Gross Ocean Product	第一产业 Primary Industry	第二产业 Secondary Industry	第三产业 Tertiary Industry
合 计 Total	**100. 0**	**5. 3**	**46. 9**	**47. 8**
天 津 Tianjin	100. 0	0. 2	66. 7	33. 1
河 北 Hebei	100. 0	4. 4	54. 0	41. 6
辽 宁 Liaoning	100. 0	13. 2	39. 5	47. 3
上 海 Shanghai	100. 0	0. 1	37. 8	62. 1
江 苏 Jiangsu	100. 0	4. 7	51. 6	43. 7
浙 江 Zhejiang	100. 0	7. 5	44. 1	48. 4
福 建 Fujian	100. 0	9. 3	40. 5	50. 2
山 东 Shandong	100. 0	7. 2	48. 6	44. 2
广 东 Guangdong	100. 0	1. 7	48. 9	49. 4
广 西 Guangxi	100. 0	18. 7	39. 7	41. 6
海 南 Hainan	100. 0	21. 6	19. 2	59. 2

2-20 沿海地区海洋及相关产业增加值
Added Values of Marine and Related Industries by Coastal Regions

单位：亿元 (100 million yuan)

地 区 Region	合 计 Total	海洋产业 Marine Industry	主要海洋产业 Major Marine Industry	海洋科研教育管理服务业 Industries of Marine Scientific Research, Education, Management and Service	海洋相关产业 Ocean-related Industries
合 计 Total	**50 045.2**	**29 264.4**	**20 829.9**	**8 434.6**	**20 780.8**
天 津 Tianjin	3 939.2	2 183.1	1 989.1	194.0	1 756.1
河 北 Hebei	1 622.0	866.1	783.3	82.8	755.9
辽 宁 Liaoning	3 391.7	2 119.8	1 658.9	460.9	1 271.9
上 海 Shanghai	5 946.3	3 508.9	2 206.7	1 302.2	2 437.4
江 苏 Jiangsu	4 722.9	2 645.2	1 965.3	679.9	2 077.8
浙 江 Zhejiang	4 947.5	2 812.1	1 964.7	847.4	2 135.4
福 建 Fujian	4 482.8	2 543.5	1 850.5	693.0	1 939.3
山 东 Shandong	8 972.1	5 259.3	3 894.6	1 364.7	3 712.9
广 东 Guangdong	10 506.6	6 317.0	3 752.2	2 564.8	4 189.6
广 西 Guangxi	761.0	469.0	394.2	74.8	292.1
海 南 Hainan	752.9	540.5	370.4	170.1	212.5

2-21 沿海地区海洋及相关产业增加值构成
Composition of the Added Values of Marine and Related Industries by Coastal Regions

单位：% (%)

地 区 Region	合 计 Total	海洋产业 Marine Industry	主要海洋产业 Major Marine Industry	海洋科研教育管理服务业 Industries of Marine Scientific Research, Education, Management and Service	海洋相关产业 Ocean-related Industries
合 计 Total	**100.0**	**58.5**	**41.6**	**16.9**	**41.5**
天 津 Tianjin	100.0	55.4	50.5	4.9	44.6
河 北 Hebei	100.0	53.4	48.3	5.1	46.6
辽 宁 Liaoning	100.0	62.5	48.9	13.6	37.5
上 海 Shanghai	100.0	59.0	37.1	21.9	41.0
江 苏 Jiangsu	100.0	56.0	41.6	14.4	44.0
浙 江 Zhejiang	100.0	56.8	39.7	17.1	43.2
福 建 Fujian	100.0	56.7	41.3	15.5	43.3
山 东 Shandong	100.0	58.6	43.4	15.2	41.4
广 东 Guangdong	100.0	60.1	35.7	24.4	39.9
广 西 Guangxi	100.0	61.6	51.8	9.8	38.4
海 南 Hainan	100.0	71.8	49.2	22.6	28.2

主要统计指标解释

1. 海洋经济 是开发、利用和保护海洋的各类产业活动以及与之相关联活动的总和。

2. 海洋生产总值 是海洋经济生产总值的简称，指按市场价格计算的沿海地区常住单位在一定时期内海洋经济活动的最终成果，是海洋产业和海洋相关产业增加值之和。

3. 海洋产业 是开发、利用和保护海洋所进行的生产和服务活动，包括海洋渔业、海洋油气业、海洋矿业、海洋盐业、海洋化工业、海洋生物医药业、海洋电力业、海水利用业、海洋船舶工业、海洋工程建筑业、海洋交通运输业、滨海旅游业等主要海洋产业以及海洋科研教育管理服务业。

4. 海洋科研教育管理服务业 是开发、利用和保护海洋过程中所进行的科研、教育、管理及服务等活动，包括海洋信息服务业、海洋环境监测预报服务、海洋保险与社会保障业、海洋科学研究、海洋技术服务业、海洋地质勘查业、海洋环境保护业、海洋教育、海洋管理、海洋社会团体与国际组织等。

5. 海洋相关产业 是指以各种投入产出为联系纽带，与主要海洋产业构成技术经济联系的上下游产业，涉及海洋农林业、海洋设备制造业、涉海产品及材料制造业、涉海建筑与安装业、海洋批发与零售业、涉海服务业等。

6. 海洋三次产业 我国的海洋三次产业划分如下：

海洋第一产业：是指海洋渔业中的海洋水产品、海洋渔业服务业，以及海洋相关产业中属于第一产业范畴的部门。

海洋第二产业：是指海洋渔业中海洋水产品加工、海洋油气业、海洋矿业、海洋盐业、海洋化工业、海洋生物医药业、海洋电力业、海水利用业、海洋船舶工业、海洋工程建筑业，以及海洋相关产业中属于第二产业范畴的部门。

海洋第三产业：是指除海洋第一、二产业以外的其他行业。第三产业包括：海洋交通运输业、滨海旅游业、海洋科研教育管理服务业，以及海洋相关产业中属于第三产业范畴的部门。

7. 海洋渔业 包括海水养殖、海洋捕捞、海洋渔业服务业和海洋水产品加工等活动。

8. 海洋油气业 是指在海洋中勘探、开采、输送、加工原油和天然气的生产活动。

9. 海洋矿业 包括海滨砂矿、海滨土砂石、海滨地热与煤矿及深海矿物等的采选活动。

10. 海洋盐业 是指利用海水生产以氯化钠为主要成分的盐产品的活动，包括采盐和盐加工。

11. 海洋船舶工业 是指以金属或非金属为主要材料，制造海洋船舶、海上固定及浮动装置的活动，以及对海洋船舶的修理及拆卸活动。

12. 海洋化工业 包括海盐化工、海水化工、海藻化工及海洋石油化工的化工产品生产活动。

13. 海洋生物医药业 是指以海洋生物为原料或提取有效成分，进行海洋药品与海洋保健品的生产加工及制造活动。

14. 海洋工程建筑业 是指在海上、海底和海岸所进行的用于海洋生产、交通、娱乐、防护等用途的建筑工程施工及其准备活动；包括海港建筑、滨海电站建筑、海岸堤坝建筑、海洋隧道桥梁建筑、海上油气田陆地终端及处理设施建造、海底线路管道和设备安装，不包括各部门、各地区的房屋建筑及房屋装修工程。

15. 海洋电力业 是指在沿海地区利用海洋能、海洋风能进行的电力生产活动。不包括沿海地区的火力发电和核力发电。

16. 海水利用业 是指对海水的直接利用和海水淡化活动，包括利用海水进行淡水生产和将海水应用于工业冷却用水和城市生活用水、消防用水等活动，不包括海水化学资源综合利用活动。

17. 海洋交通运输业 是指以船舶为主要工具从事海洋运输以及为海洋运输提供服务的活动，包括远洋旅客运输、沿海旅客运输、远洋货物运输、沿海货物运输、水上运输辅助活动、管道运输业、装卸搬运及其他运输服务活动。

18. 滨海旅游业 是指以海岸带、海岛及海洋各种自然景观、人文景观为依托的旅游经营、服务活动, 主要包括：海洋观光游览、休闲娱乐、度假住宿、体育运动等活动。

Explanatory Notes on Main Statistical Indicators

1. Marine Economy is the summation of various types of industrial activities for developing, utilizing and protecting the ocean as well as the activities associated with there.

2. Gross Ocean Product is the short form of the gross output value of ocean economy, referring to the final result of marine economic activities of the permanent units in the coastal region within a given period calculated at the market price, and the sum total of the added values of the marine industries as the ocean-related industries.

3. Marine industry refers to the production as service activities for developing, utilizing and protecting the ocean, including major marine industries such as offshore oil and gas industry, marine mining industry, marine salt industry, marine chemical industry, marine biomedicine industry, marine electric power industry, seawater utilization industry, marine shipbuilding industry, marine engineering construction industry, marine communications and transportation industry, coastal tourism etc. as well as marine scientific research, education, management and service.

4. Marine Scientific Research, Education, Management and Service refer to the activities of scientific research, education, management and service carried out in the process of developing, utilizing and protecting the ocean, including marine information service industry, marine environment monitoring and forecasting service, marine insurance and social security industry, marine scientific research, marine technological service industry, ocean geological prospecting industry, marine environmental protection industry, marine education, marine management, marine social organization and international organizations etc.

5. Ocean-Related Industry refers to the lower and upper reaches enterprises that form a technical and economic link with the major marine industries, with various inputs and outputs as ties, involving marine agriculture and forestry, marine equipment manufacturing, ocean-related building and installation industry, marine wholesale and retail industry, ocean-related service industry etc.

6. Marine Three Industries Chinese marine three industries are divided as follows:

Marine primary industry: refers to the marine aquatic products, marine fishery service industry in the marine fishery as well as the sectors belonging to the primary industry category in the ocean-related industries.

Marine secondary industry: refers to the marine aquatic products processing industry in the marine fishery, offshore oil as gas industry, marine mining industry, marine salt industry, marine chemical industry, marine biomedicine industry, marine electric power industry, seawater utilization industry, marine shipbuilding industry, marine engineering construction industry, as well as the sectors belonging to the category of secondary industry in the ocean-related industries.

Marine tertiary Industry: refers to the industries other than the marine primary and secondary industries, including marine communications and transportation industry, coastal tourism, marine scientific research, education, management and service industry as well as the sectors belonging to the category of tertiary industry in the ocean-related industries.

7. Marine Fishery includes mariculture, marine fishing, marine fishery service industry and marine aquatic products processing, etc.

8. Offshore Oil and Gas Industry refers to the production activities of exploring, exploiting, transporting and processing crude oil and natural gas in the ocean.

9. Ocean Mining Industry includes the activities of extracting and dressing beach placers, beach soil and sand, submarine geothermal energy, and coal mining and deep-sea mining, etc.

10. Marine Salt Industry refers to the activity of producing the salt products with the sodium chloride as the main component by utilizing seawater, including salt extracting and processing.

11. Shipbuilding Industry refers to the activity of building ocean vessels, offshore fixed and floating equipment with metals or non-metals as main materials as well as repairing and dismantling ocean vessels.

12. Marine Chemical Industry includes the production activities of chemical products of sea salt, seawater, sea algal and marine petroleum chemical industries.

13. Marine Biomedicine Industry refers to the production, processing and manufacturing activities of marine medicines and marine health care products by using marine organisms as raw materials or extracting useful components therefrom.

14. Marine Engineering Building Industry refers to the architectural projects construction and its preparations in the sea, at the sea bottom and seacoast for such uses as marine production, transportation, recreation, protection, etc., including constructions of seaports, coastal power stations, coastal dykes, marine tunnels and bridges, land terminals of offshore oil and gas fields as well as building of processing facilities, and installation of submarine pipelines and equipment, but not the projects of house building and renovation.

15. Marine Electric Power Industry refers to the activities of generating electric power in the coastal region by making use of ocean energies and ocean wind energy. It does not include the thermal and nuclear power generation in the coastal area.

16. Seawater Utilization Industry refers to the activities of the direct use of sea water and the seawater desalination, including those of carrying out the production of desalination and applying the seawater as water for industrial cooling, urban domestic water, water for fire fighting etc., but not the activity of the multipurpose use of seawater chemical resources.

17. Marine Communications and Transportation Industry refers to the activities of carrying out and serving the sea transportations with vessels as main vehicles, including ocean-going passagers transportation, coastal passagers transportation, ocean-going cargo transportation, coastal cargo transportation, auxiliary activities of water transportation, pipeline transportation, loading, unloading and transport as well as other transportation service activities.

18. Coastal Tourism refers to the tourist business and service activities with the backing of coastal zone, sea islands as well as a variety of natural and human landscapes of the ocean, mainly including marine sightseeing, living a life of leisure and recreation, going on vocation and getting accommodation, sports, etc.

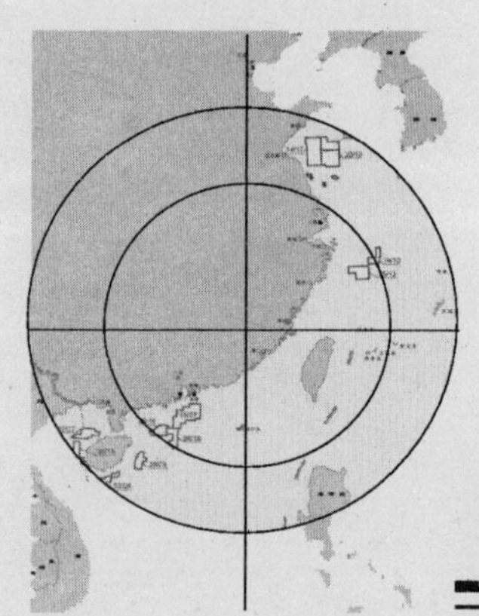

3

主要海洋产业活动

Major Marine Industrial Activities

3-1 全国海洋捕捞养殖产量
National Marine Catches and Mariculture Production

单位：吨 (t)

项　目　Item	2010	2011	2012
海水水产品产量① **Total Output of Seawater Aquatic Products**	**27 975 312**	**29 080 487**	**30 333 437**
海洋捕捞产量② **Marine Catches**	**12 035 946**	**12 419 386**	**12 671 891**
按品种分　By Species			
鱼类　Fish	8 255 051	8 639 947	8 758 466
甲壳类　Crustacea	2 043 314	2 091 282	2 207 391
贝类　Shellfish	622 104	584 078	563 422
藻类　Algae	24 636	27 362	25 728
头足类　Cephalopoda	658 309	695 251	698 909
其他　Others	432 532	381 466	417 975
海水养殖产量 **Mariculture Production**	**14 823 008**	**15 513 292**	**16 438 105**
按品种分　By Species			
鱼类　Fish	808 171	964 189	1 028 399
甲壳类　Crustacea	1 061 096	1 127 189	1 249 554
贝类　Shellfish	11 082 321	1 153 626	12 084 393
藻类　Algae	1 541 322	1 601 764	1 764 684
其他　Others	330 098	276 524	311 075

注：①海水水产品产量为海洋捕捞产量、远洋捕捞产量及海水养殖产量之和。
②海洋捕捞产量未包括远洋捕捞产量。

Note: ① The output of seawater aquatic products is the sum of the productions of marine fishing, deep-sea fishing and mariculture.
② Data for deep-sea fishing production are not included in the marine catches.

3-2 沿海地区海洋捕捞养殖产量
Marine Catches and Mariculture Production by Coastal Regions

单位：吨 (t)

地 区 Region	海洋捕捞产量 Marine Catches	远洋捕捞产量 Deep-Sea Fishing Production	海水养殖产量 Mariculture Production
全国总计 National Total	**12 671 891**	**1 223 441**	**16 438 105**
天 津 Tianjin	16 516	10 793	14 285
河 北 Hebei	252 570		382 061
辽 宁 Liaoning	1 079 288	178 376	2 635 627
上 海 Shanghai	20 387	110 198	
江 苏 Jiangsu	566 085	13 770	904 959
浙 江 Zhejiang	3 160 189	290 881	861 364
福 建 Fujian	1 927 150	212 330	3 326 595
山 东 Shandong	2 363 321	134 982	4 362 443
广 东 Guangdong	1 510 457	55 616	2 757 362
广 西 Guangxi	666 603	4 012	977 307
海 南 Hainan	1 109 325		216 102

注：远洋捕捞产量全国总计包括北京9 580吨，中农发集团202 903吨。

Note:Data for the national total deep-sea fishing production includes 9 580 tons from Beijing and 202 903 tons from the China National Agricultural Development Group Co., Ltd..

3-3　沿海地区海洋原油产量
Output of Offshore Crude Oil by Coastal Regions

单位：万吨　　(10 000 t)

地　区 Region	2010	2011	2012
合　计　Total	**4 709.98**	**4 451.97**	**4 444.79**
天　津　Tianjin	2 916.46	2 770.20	2 680.34
河　北　Hebei	221.19	229.05	237.77
辽　宁　Liaoning	13.01	10.75	14.25
上　海　Shanghai	8.80	16.10	14.68
山　东　Shandong	246.27	257.15	275.00
广　东　Guangdong	1 304.25	1 168.72	1 222.75

3-4　沿海地区海洋天然气产量
Output of Offshore Natural Gas by Coastal Regions

单位：万立方米　　($10\ 000\ m^3$)

地　区 Region	2010	2011	2012
合　计　Total	**1 108 905**	**1 214 519**	**1 228 188**
天　津　Tianjin	186 089	213 719	246 705
河　北　Hebei	40 753	50 987	55 570
辽　宁　Liaoning	3 069	2 370	1 580
上　海　Shanghai	49 742	71 789	88 044
山　东　Shandong	12 874	12 280	12 521
广　东　Guangdong	816 378	863 374	823 768

3-5 海洋原油出口量及创汇额
Export Volume of and Foreign-Exchange Earnings from Offshore Crude Oil by Coastal Regions

单位：万吨，万美元 (10 000 t, 10 000 USD)

地 区 Region	2010		2011		2012	
	出口量 Export Volume	创汇额 Foreign-Exchange Earnings	出口量 Export Volume	创汇额 Foreign-Exchange Earnings	出口量 Export Volume	创汇额 Foreign-Exchange Earnings
合 计 Total	**73.86**	**37 530**	**39.99**	**29 766**	**30.20**	**23 374**
天 津 Tianjin	48.28	23 640	25.25	19 009	18.30	13 722
广 东 Guangdong	25.58	13 890	14.74	10 757	11.90	9 652

3-6 海洋原油产量、出口量占全国原油产量、出口量比重
Proportion of Offshore Crude Oil Production and Export Volume in the National Total

年 份 Year	海洋原油产量占全国原油产量比重（%） Proportion of Offshore Crude Oil Production in the National Total (%)	海洋原油出口量占全国原油出口量比重（%） Proportion of Offshore Crude Oil Export Volume in the National Total (%)
2001	13.07	46.26
2002	14.40	54.95
2003	15.01	60.77
2004	16.16	83.37
2005	17.51	84.18
2006	17.54	89.53
2007	17.06	71.53
2008	17.96	76.46
2009	19.52	26.61
2010	23.27	24.38
2011	21.94	15.87
2012	21.42	12.43

3-7 沿海地区海洋矿业产量
Output of Marine Mining Industry by Coastal Regions

单位：吨 (t)

地 区 Region	产 量 Output		
	2010	2011	2012
合 计 Total	**34 225 357**	**42 310 427**	**43 512 901**
浙 江 Zhejiang	25 013 500	27 498 400	27 269 800
福 建 Fujian	2 287 100	2 598 700	2 645 700
山 东 Shandong	4 257 157	9 453 747	10 184 021
广 西 Guangxi	250 000	260 000	356 600
海 南 Hainan	2 417 600	2 499 580	3 056 780

3-8 沿海地区海盐产量
Sea Salt Production by Coastal Regions

单位：万吨 (10 000 t)

地 区 Region	海盐产量 Output of Sea Salt		
	2010	2011	2012
合 计 Total	**3 286.63**	**3 322.42**	**2 986.42**
天 津 Tianjin	204.40	181.00	169.95
河 北 Hebei	429.41	402.25	334.69
辽 宁 Liaoning	146.05	133.62	117.39
江 苏 Jiangsu	149.91	94.25	78.10
浙 江 Zhejiang	10.59	13.99	10.88
福 建 Fujian	29.99	48.50	27.04
山 东 Shandong	2 273.05	2 418.63	2 219.10
广 东 Guangdong	14.16	15.80	8.62
广 西 Guangxi	14.29	3.63	16.43
海 南 Hainan	14.78	10.75	4.22

3-9 沿海地区海洋化工产品产量
Output of Marine Chemical Products by Coastal Regions

单位：吨 (t)

地 区 Region	产品产量[①] Output		
	2010	2011	2012
合 计 Total	**11 499 913**	**10 890 465**	**17 450 267**
天 津 Tianjin	1 666 288	1 543 708	1 610 000
河 北 Hebei	73 960	73 960	1 046 405[②]
辽 宁 Liaoning	652 876	892 629	1 020 284
江 苏 Jiangsu	1 307 863	1 762 042	2 090 567
浙 江 Zhejiang	494 421	619 852	1 001 170
福 建 Fujian	378 987	752 636	1 462 106
山 东 Shandong	6 349 018	4 580 238	8 414 735
广 东 Guangdong	576 500[②]	665 400[②]	805 000[②]

注：①数据为沿海地区部分海洋化工企业产品汇总数据；②为中国盐业总公司数据。

Note: ① The data are collected from the products of part of the chemical enterprises in the coastal region.
② The data are from the China Salt Industry Corporation.

3-10 沿海地区海洋生物医药产品产量*
Production of the Marine Biomedicine Industry by Coastal Regions

产品名称 Name	计量单位 Unit	产品产量 Output
海参肽营养素胶囊 Sea Cucumber Peptide Nutrient Capsule	箱 box	60.00
海参酒 Sea cucumber Wine	万瓶 10 000 bottles	42.00
螺旋藻粉 Spirulina Powder	吨 t	2 425.00
螺旋藻胶囊 Spirulina Capsule	万瓶 10 000 bottles	109.64
海藻酸钠 Sodium Algrnic Acid	万吨 10 000 tons	18.24
藻酸双酯钠片 Alginic Acid Diadipose Sodium Pill	万片 10 000 pills	237.00
藻酸双酯钠（针剂） Alginic Acid Diadipose Sodium Drops	万支 10 000 bottles	34.00
鲨鱼肝油胶丸 Shark Liver Oil Pill	万粒 10 000 pellets	5 890.00
金枪鱼油胶丸 Tuna Oil Pill	万粒 10 000 pellets	1 422.00
金枪鱼油 Tuna Oil	瓶 bottle	14 709.00
卵磷脂胶丸 Lecithin Pill	万粒 10 000 pellets	253.00
鳕鱼肝油胶丸 Ling Liver Oil Pill	万粒 10 000 pellets	281.00
海藻植物胶囊 Algae Plant Capsule	吨 t	340.00
蚝贝钙片 Oyster Shell Calcium Tablets	万片 10 000 pills	8 445.48
鱼油软胶囊 Fish Oil Soft Capsule	万瓶 10 000 bottles	11.66
鱼油软胶囊 Fish Oil Soft Capsule	万盒 10 000 cases	23.71
保健品鱼油 Health Care Fish Oil	吨 t	1 530.00
鱼肝油 Cod-Liver Oil	万瓶 10 000 bottles	258.24
鱼肝油乳 Cod-Liver Oil Milk	万瓶 10 000 bottles	240.58
鱼肝脂肪油 Cod-Liver Fat Oil	吨 t	23.00

3-10 续表1 continued

产品名称 Name	计量单位 Unit	产品产量 Output
鱼肝油系列产品 Cod-Liver Oil Series Products	箱 box	2 000.00
鱼肝油系列产品 Cod-Liver Oil Series Products	吨 t	1 902.00
鲨鱼硫酸软骨素 Shark Sulphate Chondroitin	吨 t	616.95
海蛇痹宁胶囊 Sea Snake Anti-rheumatic Capsule	万粒 10 000 pellets	118.55
维生素AD胶丸 Vitamin AD Capsule	万粒 10 000 pellets	93 652.86
维生素AD滴剂 Vitamin AD Drops	万支 10 000 bottles	9 496.70
维生素E胶丸 Vitamin E Capsule	万粒 10 000 pellets	37 525.84
琼脂 Agar	吨 t	2 000.00
海珠喘息定片 Haizhu Methoxyphenamine Pill	万片 10 000 pills	22 299.00
珍珠粉末 Pearl Powder	万支 10 000 bottles	168.92
金牡感冒片 Jinmu Cold Cure Pill	万片 10 000 pills	947.27
多烯酸乙酯胶囊 Ethyl Polyenoic Acid Capsule	万瓶 10 000 bottles	1.93
多烯酸乙酯胶囊 Ethyl Polyenoic Acid Capsule	万粒 10 000 pellets	417.82
鲨力胶囊 Shali Capsule	瓶 bottle	5 000.00
氨糖美辛肠溶片 Glucosamine Indometacin Enteric-coated Tablets	万片 10 000 pills	445.45
去甲斑蝥素片 Demethylcantharidin Tablets	万片 10 000 pills	159.49
浓缩鱼油 Concentrated Fish Oil	吨 t	1 450.00
甘糖酯 Mannose Easter	万片 10 000 pills	320.00
珠珀惊风散 Zhubo Convulsions Powder	万瓶 10 000 bottles	487.84
甲壳素 Crustaceoxin	吨 t	1 600.00
脑元神软胶囊 Naoyuanshen Soft Capsule	万粒 10 000 pellets	37.92
海麟舒肝胶囊 Hailin Liver-Soothing Capsule	万粒 10 000 pellets	26.00
海威口服液 Haiwei Oral Liquid	吨 t	428.00

3-10 续表2 continued

产品名称 Name	计量单位 Unit	产品产量 Output
伊可新 Vitamin A and D Drops	吨 t	170.00
珍珠胎囊口服液 Pearl Capsule Oral Liquid	万支 10 000 bottles	415.00
麝珠明目滴眼液 Shezhu Eye-brightening Drops	万瓶 10 000 bottles	221.00
卡拉胶 Carrageenan	吨 t	7 181.00
角鲨烯 Houndfish Alkene	万瓶 10 000 bottles	2.08
角鲨烯胶囊 Houndfish Alkene Capsule	万粒 10 000 pellets	368.18
角鲨烯原料 Squalene Raw Material	吨 t	4.00
鱼蛋白粉 Fish Albumen Powder	吨 t	14.08
胶原蛋白 Collagen Albumen	吨 t	189.12
鱼油微囊 Fish Oil Capsule	吨 t	6 938.00
鱼胶原蛋白肽 Fish Collagen Peptide	万粒 10 000 pellets	139.81
海狗油软胶囊 Fur Seal Oil Soft Capsule	万瓶 10 000 bottles	2.66
八宝惊风散 Babao Infantile Convulsions Powder	万盒 10 000 cases	684.50
甘露醇 Mannitol	吨 t	64.00
碘 Iodine	吨 t	2 201.00
润科DHA粉剂、油剂 Runke DHA Powder, oil	吨 t	1 500.00
润科ARA粉剂、油剂 Runke ARA Powder,oil	吨 t	900.00
南海岸鳗钙系列产品 South Coast Eel-calcium	万盒 10 000 cases	129.88
伤科接骨片 Traumatologic Osteopathic Tablets	吨 t	316.08
肤疹宁软膏 Rash Cream	万支 10 000 bottles	4.00
康体通胶囊 Kangtitong Capsule	万粒 10 000 pellets	5 000.00
鲨鱼软骨罐头 Shark Cartilage Can	罐 can	30 000.00

注：数据为沿海地区部分海洋生物医药产品汇总数据。

Note: The data are collected from the products of part of the marine biomedicine enterprises in the coastal region.

3-11 沿海地区海洋修造船完工量
Production of the Marine Shipbuilding Industry by Coastal Regions

部门和地区 Sector and Region	修船完工量（艘） Ships Repaired (unit)	造船完工量 Ships Built	
		艘 (unit)	万综合吨 Comprehensive Tonnages (10 000 t)
总 计 Total	**13 898**	**3 495**	**5 701.60**
其 中: Including:			
中船工业集团公司 CSSC	541	158	1 352.78
中船重工集团公司 CSIC	539	106	1014.40
按地区分: By Regions:			
天 津 Tianjin	172	36	60.66
河 北 Hebei	240	10	67.22
辽 宁 Liaoning	640	114	1 270.00 ①
上 海 Shanghai	1 258	98	1 045.50
江 苏 Jiangsu	3 047	709	2 154.58
浙 江 Zhejiang	3 511	1 161	307.06
福 建 Fujian	2 620	836	177.00
山 东 Shandong	1 918	354	310.45 ①
广 东 Guangdong	395	61	308.07
广 西 Guangxi	79	15	0.23
海 南 Hainan	18	101	0.83

注：①表示单位为万载重吨。

Note: ①the unit is 10 thonsand dead weifht tonnage.

3-12 沿海地区海洋货物运输量和周转量
Volume of Maritime Goods Transported and Turnover by Coastal Regions

单位：万吨，亿吨千米 (10 000 t, 100 million t-km)

地 区 Region	货运量 Volume of Goods Transported	沿 海 Coastal	远 洋 Oceangoing	货物周转量 Volume of Goods Turnover	沿 海 Coastal	远 洋 Oceangoing
全国总计 National Total	**217 404**	**152 516**	**64 888**	**72 396**	**19 673**	**52 723**
天 津 Tianjin	10 332	4 161	6 171	7 012	768	6 245
河 北 Hebei	2 590	2 590		510	510	
辽 宁 Liaoning	12 615	5 642	6 973	7 483	776	6 707
上 海 Shanghai	47 640	30 149	17 491	20 018	3 932	16 087
江 苏 Jiangsu	17 632	12 110	5 522	5 229	1 145	4 085
浙 江 Zhejiang	45 900	43 455	2 445	6 956	5 152	1 805
福 建 Fujian	18 397	16 318	2 079	2 912	2 258	654
山 东 Shandong	9 356	7 711	1 645	2 239	814	1 425
广 东 Guangdong	28 530	17 960	10 570	6 344	2 761	3 583
广 西 Guangxi	4 738	4 297	441	787	768	20
海 南 Hainan	9 121	8 123	998	1 373	790	583
其 他* Others	10 553	0	10 553	11 530	0	11 530

注：其他数据为中国远洋运输（集团）总公司完成的远洋运输量（表3-15同）。

Note: Other data are those of the freight volume of ocean transportation accomplished by the China National Ocean Shipping Corporation (Same as Table 3-15).

3-13 沿海地区海洋旅客运输量和周转量
Volume of Maritime Passenger Traffic and Turnover by Coastal Regions

单位：万人，亿人千米　　　　(10 000 persons, 100 million person-km)

地 区 Region	客运量 Passenger Traffic	沿 海 Coastal	远 洋 Oceangoing	旅客周转量 Passenger Turnover Volume	沿 海 Coastal	远 洋 Oceangoing
全国总计 National Total	**10 610**	**9 641**	**970**	**42.05**	**30.23**	**11.82**
天 津 Tianjin	1	0	1	0.15	0.00	0.15
辽 宁 Liaoning	588	572	16	7.50	6.78	0.73
上 海 Shanghai	353	353		0.99	0.99	
江 苏 Jiangsu	23	23		0.63	0.63	
浙 江 Zhejiang	2 589	2 589		5.18	5.18	
福 建 Fujian	1 449	1 358	91	2.27	1.64	0.63
山 东 Shandong	1 760	1 657	103	12.17	7.42	4.75
广 东 Guangdong	2 367	1 608	759	9.18	3.61	5.57
广 西 Guangxi	179	179		0.92	0.92	
海 南 Hainan	1 301	1 301		3.06	3.06	

3-14 沿海港口客货吞吐量
Volume of Passenger and Freight Handled at Coastal Seaports

单位：万吨，万人次 (10 000 t, 10 000 person-times)

地 区 Region	货物吞吐量 Cargo Handled	# 外 贸 Foreign Trade	旅客吞吐量 Passenger Leaving & Arriving	# 离 港 Leaving
合 计 Total	**687 975**	**278 552**	**7 879**	**3 994**
天 津 Tianjin	47 697	24 326	29	15
河 北 Hebei	76 234	21 084	4	2
辽 宁 Liaoning	88 502	17 601	662	333
上 海 Shanghai	63 740	35 825	130	66
江 苏 Jiangsu	20 504	9 905	12	6
浙 江 Zhejiang	92 760	36 516	834	414
福 建 Fujian	41 359	16 695	1 115	557
山 东 Shandong	106 655	58 925	1 313	656
广 东 Guangdong	121 265	44 952	2 457	1 278
广 西 Guangxi	17 438	10 517	23	11
海 南 Hainan	11 819	2 207	1 297	654

3-15 沿海地区水路国际标准集装箱运量
Volume of International Standardized Containers Traffic by Coastal Regions

单位：万标准箱，万吨 (10 000 TEU, 10 000 t)

地 区 Region	2010		2011		2012	
	箱 数 Number of Containers	重 量 Weight	箱 数 Number of Containers	重 量 Weight	箱 数 Number of Containers	重 量 Weight
合 计 Total	**3 659**	**40 256**	**4 074**	**49 348**	**4 357**	**52 301**
天 津 Tianjin	12	177	8	85	6	62
河 北 Hebei	1	24	1	36	1	20
辽 宁 Liaoning	44	444	45	521	42	473
上 海 Shanghai	1 623	20 561	1 759	22 981	2 010	26 378
江 苏 Jiangsu	392	3 283	462	3 961	479	4 063
浙 江 Zhejiang	102	1 515	168	2 781	220	2 948
福 建 Fujian	231	3 902	268	4 673	302	4 373
山 东 Shandong	166	2 227	123	2 345	139	1 927
广 东 Guangdong	913	6 812	887	7 342	786	6 618
广 西 Guangxi	77	992	181	2 367	210	2 598
海 南 Hainan	100	317	172	2 256	163	2 839

注：国际标准集装箱运量数据包含内河集装箱运量。

Note: The data in this table include the volume of containers transported in the inland rivers.

3-16 沿海港口国际标准集装箱吞吐量
International Standardized Containers Handled at Coastal Seaports

单位：万标准箱，万吨 (10 000TEU, 10 000 t)

地　区 Region	2010		2011		2012	
	箱　数 Number of Containers	重　量 Weight	箱　数 Number of Containers	重　量 Weight	箱　数 Number of Containers	重　量 Weight
合　计 Total	**13 145**	**137 076**	**14 632**	**157 969**	**15 797**	**175 986**
天　津 Tianjin	1 009	10 916	1 159	11 929	1 230	13 442
河　北 Hebei	62	997	77	1 251	90	1 351
辽　宁 Liaoning	969	15 666	1 200	19 115	1 514	24 733
上　海 Shanghai	2 907	27 992	3 174	31 220	3 253	32 480
江　苏 Jiangsu	391	3 810	488	4 626	504	5 012
浙　江 Zhejiang	1 404	12 276	1 584	15 747	1 759	17 811
福　建 Fujian	867	10 743	970	12 099	1 073	13 308
山　东 Shandong	1 531	15 189	1 691	17 738	1 899	19 912
广　东 Guangdong	3 868	37 413	4 103	41 315	4 256	44 495
广　西 Guangxi	56	926	74	1 165	82	1 358
海　南 Hainan	82	1 147	112	1 764	137	2 084

3-17 沿海城市国内旅游人数
Domestic Visitors by Coastal Cities

单位：万人次 (10 000 person-times)

城　市	City	2009	2010	2011
合　计	**Total**	**91 807**	**107 019**	**135 467**
天　津	**Tianjin**	5 537	6 118	10 605
河　北	**Hebei**	3 362	3 968	4 790
唐　山	Tangshan	1 226	1 532	2 001
秦皇岛	Qinhuangdao	1 638	1 861	2 101
沧　州	Cangzhou	498	575	688
辽　宁	**Liaoning**	9 800	11 655	13 585
大　连	Dalian	3 412	3 777	4 261
丹　东	Dandong	1 897	2 248	2 646
锦　州	Jinzhou	1 191	1 442	1 704
营　口	Yingkou	820	1 074	1 279
盘　锦	Panjin	1 270	1 654	1 959
葫芦岛	Huludao	1 211	1 460	1 736
上　海	**Shanghai**	12 361	21 463	23 079
江　苏	**Jiangsu**	3 655	4 255	5 090
南　通	Nantong	1 483	1 757	2 109
连云港	Lianyungang	1 210	1 393	1 656
盐　城	Yancheng	962	1 105	1 325
浙　江	**Zhejiang**	21 948	26 371	30 547
杭　州	Hangzhou	5 094	6 305	7 181
宁　波	Ningbo	3 962	4 624	5 181
温　州	Wenzhou	2 931	3 537	4 123
嘉　兴	Jiaxing	2 492	3 070	3 536
绍　兴	Shaoxing	2 851	3 436	4 128
舟　山	Zhoushan	1 731	2 113	2 433
台　州	Taizhou	2 887	3 286	3 965
福　建	**Fujian**	6 938	8 262	9 280
福　州	Fuzhou	1 938	2 275	2 680
厦　门	Xiamen	1 788	2 178	2 569
莆　田	Putian	632	775	1 165
泉　州	Quanzhou	1 171	1 351	1 083
漳　州	Zhangzhou	835	1 002	951
宁　德	Ningde	574	681	832

3-17 续表 continued

城 市 City		2009	2010	2011
山 东	**Shandong**	13 580	16 055	18 807
青 岛	Qingdao	3 903	4 397	4 956
东 营	Dongying	515	637	775
烟 台	Yantai	2 763	3 272	3 863
潍 坊	Weifang	2 313	2 946	3 603
威 海	Weihai	1 839	2 112	2 372
日 照	Rizhao	1 724	2 031	2 426
滨 州	Binzhou	523	661	812
广 东	**Guangdong**	11 696	13 612	15 538
广 州	Guangzhou	3 286	3 692	3 816
深 圳	Shenzhen	1 944	2 265	2 628
珠 海	Zhuhai	911	1 055	1 215
汕 头	Shantou	668	769	890
江 门	Jiangmen	746	872	1 013
湛 江	Zhanjiang	462	602	909
茂 名	Maoming	201	305	360
惠 州	Huizhou	799	913	1 014
汕 尾	Shanwei	239	327	439
阳 江	Yangjiang	273	305	456
东 莞	Dongguan	1 191	1 289	1 400
中 山	Zhongshan	504	540	605
潮 州	Chaozhou	274	317	361
揭 阳	Jieyang	198	361	432
广 西	**Guangxi**	1 630	1 958	2 347
北 海	Beihai	810	938	1 101
防城港	Fangchenggang	418	550	676
钦 州	Qinzhou	402	469	570
海 南	**Hainan**	1 299	1 564	1 799
海 口	Haikou	662	723	831
三 亚	Sanya	637	841	968

注：数据来源于《中国区域经济统计年鉴》（2012）。

Note：The data come from *China Statistical Yearbook For Regional Economy (2012)*.

3-18 主要沿海城市国际旅游（外汇）收入
(Foreign Exchange) Earnings from International Tourism in Major Coastal Cities

单位：万美元 (10 000 USD)

城市	City	2010	2011	2012
天 津	Tianjin	141 951	175 553	222 641
秦皇岛	Qinhuangdao	12 022	13 770	19 454
大 连	Dalian	80 386	80 519	87 349
上 海	Shanghai	634 092	575 118	549 323
南 通	Nantong	36 066	39 916	42 995
连云港	Lianyungang	10 747	12 869	14 434
杭 州	Hangzhou	169 008	195 710	220 165
宁 波	Ningbo	59 066	65 472	73 428
温 州	Wenzhou	21 115	25 602	31 887
福 州	Fuzhou	84 300	102 854	110 817
厦 门	Xiamen	108 552	129 901	157 728
泉 州	Quanzhou	66 737	79 661	90 446
漳 州	Zhangzhou	15 455	18 781	22 878
青 岛	Qingdao	60 103	68 933	82 459
烟 台	Yantai	37 707	46 816	48 146
威 海	Weihai	19 151	21 855	25 283
广 州	Guangzhou	466 127	485 306	514 458
深 圳	Shenzhen	315 896	374 474	432 882
珠 海	Zhuhai	122 339	106 685	95 045
汕 头	Shantou	5 016	5 071	5 175
湛 江	Zhanjiang	2 716	3 630	4 816
中 山	Zhongshan	27 591	24 717	21 979
北 海	Beihai	2 173	2 574	3 429
海 口	Haikou	3 756	3 845	4 474
三 亚	Sanya	24 504	31 259	26 565

3-19 主要沿海城市接待入境旅游者人数
Number of Inbound Tourists Received by in Major Coastal Cities

单位：人次 (person-time)

城 市 City		2010	2011	2012
天 津	Tianjin	1 660 682	730 615	737 481
秦皇岛	Qinhuangdao	242 337	264 372	286 401
大 连	Dalian	1 166 020	1 170 035	1 284 176
上 海	Shanghai	7 337 216	6 686 144	6 512 347
南 通	Nantong	355 133	404 852	440 788
连云港	Lianyungang	116 663	132 289	144 684
杭 州	Hangzhou	2 757 147	3 063 140	3 311 225
宁 波	Ningbo	951 680	1 073 872	1 162 088
温 州	Wenzhou	391 587	470 504	575 397
福 州	Fuzhou	698 607	761 665	851 328
厦 门	Xiamen	1 551 865	1 799 205	2 124 163
泉 州	Quanzhou	770 457	846 389	951 424
漳 州	Zhangzhou	247 469	293 728	330 669
青 岛	Qingdao	1 080 511	1 156 391	1 270 076
烟 台	Yantai	472 023	548 533	530 184
威 海	Weihai	372 646	415 114	456 594
广 州	Guangzhou	8 147 900	7 786 900	7 866 031
深 圳	Shenzhen	10 206 000	11 045 500	12 064 451
珠 海	Zhuhai	3 251 400	3 208 300	2 975 791
汕 头	Shantou	133 800	140 700	147 800
湛 江	Zhanjiang	103 200	142 300	147 400
中 山	Zhongshan	480 600	608 000	558 573
北 海	Beihai	73 008	83 073	98 759
海 口	Haikou	132 877	146 808	179 741
三 亚	Sanya	415 939	528 942	481 437

3-20 主要沿海城市接待入境旅游者情况

Breakdown of Inbound Tourists Received by Major Coastal Cities

单位：人次，人天 (person-time, night)

城市 City	外国人 Foreigners		香港同胞 Hong Kong	
	人次数 Arrivals	人天数 Nights	人次数 Arrivals	人天数 Nights
天　津 Tianjin	637 106	8 601 565	46 254	890 551
秦皇岛 Qinhuangdao	268 480	1 165 116	8 119	28 823
大　连 Dalian	1 125 216	3 750 188	74 088	247 757
上　海 Shanghai	5 396 384	17 872 209	450 548	1 404 776
南　通 Nantong	391 157	2 135 448	18 734	120 623
连云港 Lianyungang	116 952	700 918	9 796	66 810
杭　州 Hangzhou	2 298 763	6 943 834	384 642	1 123 145
宁　波 Ningbo	631 033	1 883 598	186 801	515 853
温　州 Wenzhou	440 546	1 237 864	54 659	120 270
福　州 Fuzhou	477 274	3 462 820	101 037	710 199
厦　门 Xiamen	761 100	3 820 501	220 364	913 853
泉　州 Quanzhou	159 515	969 289	519 427	2 803 769
漳　州 Zhangzhou	77 593	318 864	96 710	362 650
青　岛 Qingdao	877 774	2 448 148	194 059	680 304
烟　台 Yantai	416 891	1 923 953	32 259	88 132
威　海 Weihai	428 464	1 150 846	3 448	7 927
广　州 Guangzhou	2 892 049	7 562 598	3 934 448	8 091 321
深　圳 Shenzhen	1 691 045	3 780 960	9 863 344	19 667 829
珠　海 Zhuhai	538 282	1 214 571	1 130 728	1 929 405
汕　头 Shantou	93 200	185 500	44 700	83 200
湛　江 Zhanjiang	73 600	152 400	60 400	138 700
中　山 Zhongshan	115 830	458 907	314 067	750 157
北　海 Beihai	53 205	95 299	31 130	50 099
海　口 Haikou	82 609	135 473	26 074	41 704
三　亚 Sanya	367 819	1 291 700	53 377	112 473

3-20 续表 continued

城市 City	澳门同胞 Macao		台湾同胞 Taiwan Province	
	人次数 Arrivals	人天数 Nights	人次数 Arrivals	人天数 Nights
天津 Tianjin	2 698	58 745	51 423	884 214
秦皇岛 Qinhuangdao	719	2 581	9 083	30 742
大连 Dalian	2 156	7 695	82 716	276 384
上海 Shanghai	18 176	63 598	647 239	2 384 988
南通 Nantong	2 164	9 944	28 733	188 652
连云港 Lianyungang	5 049	18 755	12 887	120 176
杭州 Hangzhou	42 511	71 601	585 309	1 626 130
宁波 Ningbo	61 632	170 582	282 622	705 084
温州 Wenzhou	17 120	27 749	63 072	139 461
福州 Fuzhou	7 135	48 295	265 882	993 725
厦门 Xiamen	7 762	35 231	1 134 937	3 352 235
泉州 Quanzhou	51 472	244 142	221 010	877 123
漳州 Zhangzhou	9 201	27 194	147 165	424 163
青岛 Qingdao	46 045	123 043	152 198	479 066
烟台 Yantai	16 941	45 510	64 093	189 521
威海 Weihai	879	2 092	23 803	65 871
广州 Guangzhou	498 683	1 034 483	540 851	1 356 567
深圳 Shenzhen	54 835	98 481	455 227	1 174 177
珠海 Zhuhai	712 063	1 425 316	594 718	1 369 525
汕头 Shantou	700	1 300	9 200	18 100
湛江 Zhanjiang	3 700	8 100	9 700	20 100
中山 Zhongshan	70 470	144 870	58 206	140 348
北海 Beihai	4 728	8 222	9 696	17 659
海口 Haikou	1 318	2 143	69 740	88 547
三亚 Sanya	3 413	7 323	56 828	107 034

主要统计指标解释

1. 海洋捕捞产量 凡是从海洋里捕捞的天然生长的水产品产量为捕捞产量。

2. 海水养殖产量 凡是从人工投放苗种或天然纳苗并进行人工饲养管理的海水养殖水域中捕捞的水产品产量为海水养殖产量。

3. 远洋捕捞产量 由各远洋渔业企业和各生产单位按我国远洋渔业项目管理办法组织的远洋渔船（队）在非我国管辖海域（外国专属经济区水域或公海）捕捞的水产品产量。中外合资、合作渔船捕捞的水产品只统计按协议应属于中方所有的部分。

4. 原油产量 是按净原油量来计算的，能直接用于销售和生产自用的原油量。目前海洋石油系统原油产量计算方法采用倒算法。

原油产量=销售量+期末库存量-期初库存量+海上平台及陆地终端处理厂自用量。

5. 天然气产量 指进入集输管网的销售量和就地利用的全部气量。

天然气产量=外输（销）量+企业自用量

6. 海洋原油出口量 指销往国外的产品数量。

7. 海洋原油出口创汇额 指产品销往国外的归中方所有的全部外汇收入。以美元或万美元表示。

8. 造船综合吨 等于以计量单位载重吨和满载排水量吨的民用船舶的吨位数之和。

9. 货运量 指经船舶实际运送的货物重量，按到达量统计。

10. 货物周转量 指实际运送的货物与其运送距离的乘积。

11. 集装箱运量 既包含货重，也包含箱重。箱重系指承运租用的空箱重量凡有运费收入的空箱，其重量应统计为运量，按空箱 1 吨为货运量 1 吨计算；若无收入，所承运的空箱一律不作运量统计。

12. 旅客周转量 指实际运送的旅客人数与其运送距离的乘积。

13. 国际旅游外汇收入 入境游客在中国（大陆）境内旅行、游览过程中用于交通、参观游览、住宿、餐饮、购物、娱乐等的全部花费。

14. 接待人次数 指报告期内我国接待游客人数。游客按出游地分为入境游客和国内游客，按出游时间分为旅游者（过夜游客）和一日游游客（不过夜游客）。

15. 接待人天数 指过夜旅游者的停留天数。

16. 外国人 指外国国籍的人，加入外籍的中国血统华人也计入外国人。

17. 港澳台同胞 指居住在我国香港特别行政区、澳门特别行政区和台湾省的中国同胞。

Explanatory Notes on Main Statistical Indicators

1. Marine Catches refers to the output of the naturally growing aquatic products caught from the sea.

2. Mariculture Production refers to the output of aquatic products whose young are artificially released or naturally collected, and raised and managed artificially, and which are caught from the waters of mariculture.

3. Deep-Sea Fishing Production refers to the output of aquatic products caught in the non-Chinese jurisdictional sea areas (foreign EEE or high sea) by the distant fishing vessels (fleet) organized by various distant fishing businesses and production units according to the management measures of the China distant fishing projects. The aquatic products caught by the Chinese-foreign joint ventures' and cooperative fishing vessels are counted only for the part owned by the Chinese side according to the agreement.

4. Output of Crude Oil is calculated on the basis of the net amount of crude oil, i.e., the amount of crude oil that may be directly used for sale and for the production itself.

Output of crude oil = Volume of sales + Reserves at the end of the period - Reserves at the beginning of the period +Amount for self-use on the platforms and in the terminal processing plants on land.

5. Output of Natural Gas refers to the total gas volume of the sales volume entering the oil collecting and transport pipeline network and that used locally.

Output of natural gas = Volume of sales or transport to other areas + Volume used by the enterprise itself

6. Export Volume of Offshore Crude Oil refers to the amount of products for sale abroad.

7. Foreign Exchange Earnings of Offshore Crude Oil refer to the total foreign exchange income from oil (gas) products for sale abroad which is owned by the Chinese side.

8. Comprehensive Tonnages of Shipbuilding refers to the sum of tonnage of civilian vessels with the deadweight capacity and full-load displacement as measured.

9. Freight Traffic refers to the weight of cargoes actually transported by vessels, which is counted according to the volume of arrival.

10. Cargoes Turnover Volume refers to the product of the actually transported cargoes and the transport distance.

11. Freight Volume of Containers includes the weight of both cargoes and container boxes. The weight of container boxes refers to the weight of empty containers rented for transport or having freight income and should be included in the freight volume, one ton of empty boxes equalling to one ton of freight volume. The empty boxes which have no income for transportation are not included in the freight volume.

12. Passenger Turnover Volume refers to the product of the number of passengers actually transported and the shipping distance.

13. International Tourism (Foreign Exchange) Receipts refer to the total expenditure made by inbound visitors within the territory of China (the mainland) in their course of travel on transport, tours and sightseeing, lodging, food and beverage, shopping, entertainment, etc.

14. Number of Person-Times Received refers to the number of visitors received by China in the period reported. Visitors are divided into inbound visitors and domestic visitors by origin of the travel, and tourists (overnight visitors) and same-day visitors by their length of stay.

15. Number of the Days of Stay refers to the number of the days of stay of tourists.

16. Foreigners refer to the people with foreign nationality, including foreign nationals of Chinese descent.

17. Compatriots from Hong Kong, Macao and Taiwan Province refer to the Chinese compatriots living in the Hong Kong Special Administrative Region, the Macao Special Administrative Region and Taiwan Province.

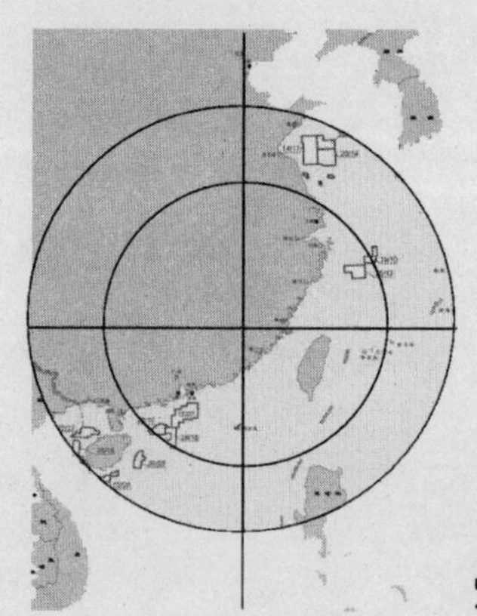

4

主要海洋产业生产能力

Production Capacity of Major Marine Industries

4-1 沿海地区渔港情况
Fishing Ports in the Coastal Regions

单位：个　　　　(unit)

地　区 Region		渔港合计 The Total of the Fishing Ports	中心渔港 Central Fishing Port	一级渔港 Grade 1 Fishing Port
合　计	**Total**	**137**	**60**	**77**
天　津	Tianjin			
河　北	Hebei	7	3	4
辽　宁	Liaoning	17	6	11
上　海	Shanghai			
江　苏	Jiangsu	10	6	4
浙　江	Zhejiang	24	9	15
福　建	Fujian	18	7	11
山　东	Shandong	24	11	13
广　东	Guangdong	19	8	11
广　西	Guangxi	8	4	4
海　南	Hainan	10	6	4

资料来源：《2013中国渔业统计年鉴》。
Data Source: *China Fishery Statistical Yearbook 2013*.

4-2 沿海地区海水养殖面积
Mariculture Area by Coastal Regions

单位：公顷 (hm²)

地 区 Region	海水养殖面积 Mariculture Area
合 计 **Total**	**2 180 927**
天 津 Tianjin	3 992
河 北 Hebei	134 682
辽 宁 Liaoning	813 035
上 海 Shanghai	
江 苏 Jiangsu	199 352
浙 江 Zhejiang	89 747
福 建 Fujian	145 486
山 东 Shandong	523 705
广 东 Guangdong	201 834
广 西 Guangxi	53 249
海 南 Hainan	15 845

4-3 海洋油田生产井情况
Survey of Offshore Oilfield Production Wells

单位：口 (well)

地区 Region	合计 Total	采油井 Oil Wells	采气井 Gas Wells	注水井 Injection Wells	其他井 Others
合计 Total	**5 632**	**4 219**	**264**	**1 138**	**11**
天津 Tianjin	3 206	2 392	130	684	
河北 Hebei	919	707	2	210	
辽宁 Liaoning	238	183	6	38	11
上海 Shanghai	43	13	30		
山东 Shandong	629	439	8	182	
广东 Guangdong	597	485	88	24	

4-4 海洋石油勘探情况
Work Volume of Offshore Oil Exploration

地区 Region	地震测线 Seismic Line		钻井（口） Drilling (well)	
	二维（千米） Two Dimensions (km)	三维（平方千米） Three Dimensions (km^2)	预探井 Wildcat Wells	评价井 Appraisal Wells
合计 Total	**17 103**	**27 072**	**77**	**96**
天津 Tianjin		7 215	19	49
河北 Hebei			15	14
辽宁 Liaoning	960		5	
上海 Shanghai	6 461	4 218	3	4
山东 Shandong		339	10	12
广东 Guangdong	9 682	11 925	22	17
其中：合作 Including: Cooperative		3 375	3	

4-5 沿海地区盐田面积和海盐生产能力
Salt Pan Area and Sea Salt Production Capacity by Coastal Regions

地 区 Region	盐田总面积（公顷） Total Area of Salt Pan (hm^2)		生产面积（公顷） Production Area (hm^2)		年末海盐生产能力（万吨） Year-End Capacity of Sea Salt Production (10 000 t)	
	2011	2012	2011	2012	2011	2012
合 计 Total	**455 433**	**441 482**	**359 264**	**323 196**	**3 757.25**	**3 641.41**
天 津 Tianjin	28 123	27 258	27 461	26 470	166.56	166.56
河 北 Hebei	71 456	84 350	63 930	63 951	461.10	465.10
辽 宁 Liaoning	33 059	34 632	28 415	29 091	207.22	179.17
江 苏 Jiangsu	71 447	60 209	42 365	19 179	110.13	87.81
浙 江 Zhejiang	2 790	2 754	2 303	2 247	14.52	13.16
福 建 Fujian	5 607	6 316	4 893	4 884	53.48	37.94
山 东 Shandong	225 871	209 753	179 518	167 316	2 700.00	2 648.24
广 东 Guangdong	10 052	10 052	6 014	6 014	18.53	18.62
广 西 Guangxi	3 373	2 504	1 557	1 236	7.39	6.49
海 南 Hainan	3 655	3 654	2 808	2 808	18.32	18.32

注：本表数据（除海南省）为沿海城市合计数，海南省为沿海地带合计数。
Note: The data in this table (except that for Hainan Province) are for the total number of coastal cities and these for Hainan Province are for the total number of coastal zones.

4-6 沿海地区风能发电能力
Wind Power Production by Coastal Regions

单位：万千瓦 (10 000 kW)

地 区 Region	风能年发电能力 Annual Wind Power Generation Capacity	
	2011	2012
合 计 Total	**1 121.57**	**1 352.94**
天 津 Tianjin	24.35	27.80
辽 宁 Liaoning	114.45	142.94
河 北 Hebei	20.55	22.05
上 海 Shanghai	31.80	35.20
江 苏 Jiangsu	195.51	230.55
浙 江 Zhejiang	35.22	46.67
福 建 Fujian	102.57	128.57
山 东 Shandong	441.16	534.06
广 东 Guangdong	130.04	154.38
广 西 Guangxi	0.25	0.25
海 南 Hainan	25.67	30.47

注：本表数据（除海南省）为沿海城市合计数，海南省为沿海地带合计数。
Note: The data in this table (except that for Hainan Province) are for the total number of coastal cities and these for Hainan Province are for the total number of coastal zones.

4-7　主要潮汐电站分布情况
Distribution of Major Tidal Power Stations

电站名称 Name	运行情况 Status of Operation	装机容量（千瓦） Capacity (kW)
江厦潮汐试验电站	1970年开始建造，1980年投入使用，运行至今。	3 900
Jiangxia Experimental Tidal Power Station	It began construction in 1970, was put into use in 1980, and has been in operation so far.	
海山潮汐电站	1972年开始建造，1975年投入使用，停止运行。	
Haishan Tidal Power Station	It began construction in 1972, was put into use in 1975, and stopped power generation .	
岳浦潮汐电站	1970年开始建造，1978年停止运行。	
Yuepu Tidal Power Station	It began construction in 1970, and stopped power generation in 1978.	
白沙口潮汐电站	1970年开始建造，1978年投入使用，2010年停止运行。	
Baishakou Tidal Power Station	It began construction in 1970, was put into use in 1978, and stopped power generation in 2010.	

4-8 主要海上活动船舶
Major Vessels Operating on the Sea

类 别 Type	艘数（艘） Number of Vessels (unit)	总吨（万吨位） grt (10 000 t)	净载重量（万吨） Net Weight Tonnage (10 000 t)	载客量（客位） Passenger Spaces (seat)	总功率（千瓦） Total Power (kW)
一 、海洋生产用船 Vessels for Marine Production					
海洋渔业船舶 Marine Fishing Vessels	280 450	771.43			15 950 851
远洋渔船 Ocean-going Fishing Vessels	1 793				1 111 754
海洋油气船舶 Offshore Oil and Gas Vessels	174	98.00	46.00	10 090	911 201
钻井平台 Drilling Vessels	40	37.00	11.00	5 412	263 072
物探船 Physical Exploration Vessels	9	3.00	1.00	469	42 601
其 他 Others	125	59.00	34.00	4 209	605 528
海洋运输船舶 Maritime Transport Vessels	12 692	8 101.60	13 039.80	208 547	32 938 245
二、海洋科研用船 Vessels for Marine Scientific Research					
海洋地质勘探船 Marine Geological Survey Vessels	6	1.19	0.36	245	23 382
海洋调查船 Marine Research Vessels					
中国科学院 Chinese Acadamy of Sciences	7	1.37	0.46	295	30 444
国家海洋局 State Oceanic Administration	4	2.60	1.35	318	25 012
三、海洋执法用船 Vessels for Marine Law Enforcement					
海监船* Marine Surveillance Vessels	562				
渔政执法船 Fishing law Enforcement	2 300	6.83			375 659

注：*包含船、艇数据。

Note: * include the data of ships and boats.

4-9 沿海规模以上港口生产用码头泊位

Berths for Productive Use at the coastal Seaports above Designed Size

单位：米, 个 (m, unit)

港口	Seaport	码头长度 Length of Quay Line	泊位个数 Number of Berths	#万吨级 10 000 Tonnage Class
合计	**Total**	**647 541**	**4 811**	**1 453**
丹东	Dandong	6 407	38	21
大连	Dalian	36 872	206	93
营口	Yingkou	16 164	75	49
锦州	Jinzhou	5 375	20	18
秦皇岛	Qinhuangdao	14 750	66	42
黄骅	Huanghua	5 570	25	19
唐山	Tangshan	16 628	63	60
天津	Tianjin	32 630	148	101
烟台	Yantai	17 020	85	56
威海	Weihai	3 949	20	12
青岛	Qingdao	20 944	79	63
日照	Rizhao	12 985	52	45
上海	Shanghai	74 459	612	152
连云港	Lianyungang	11 236	56	41
嘉兴	Jiaxing	7 463	52	25
宁波-舟山	Ningbo-Zhoushan	75 735	601	137
台州	Taizhou	11 325	174	7
温州	Wenzhou	16 516	237	16

4-9 续表 continued

港口 Seaport		码头长度 Length of Quay Line	泊位个数 Number of Berths	#万吨级 10 000 Tonnage Class
福　州	Fuzhou	22 199	178	47
莆　田	Putian	4 772	45	7
泉　州	Quanzhou	13 802	104	19
厦　门	Xiamen	25 358	145	64
汕　头	Shantou	9 444	86	18
汕　尾	Shanwei	1 286	14	1
惠　州	Huizhou	6 957	40	18
深　圳	Shenzhen	30 055	160	68
虎　门	humen	12 992	104	22
广　州	Guangzhou	45 255	493	66
中　山	Zhongshan	3 487	61	0
珠　海	Zhuhai	13 479	129	18
江　门	Jiangmen	9 551	147	2
阳　江	Yangjiang	2 232	10	9
茂　名	Maoming	2 428	18	9
湛　江	Zhanjiang	15 757	153	31
北部湾港	Beibuwan	30 624	240	65
海　口	Haikou	4 884	36	10
洋　浦	Yangpu	5 197	29	14
八　所	Basuo	1 754	10	8

4-10 沿海地区星级饭店基本情况
Star Grade Hotels and Occupancies by Coastal Regions

地 区 Region	饭店数（座） Number of Hotels (unit)	客房数（间） Number of Rooms (unit)	床位数（张） Number of Beds (unit)	客房出租率 （%） Room Occupancy (%)
合 计 Total	**5 309**	**754 199**	**1 356 182**	**57.77**
天 津 Tianjin	103	17 831	28 975	49.58
河 北 Hebei	404	51 566	95 183	50.79
辽 宁 Liaoning	399	64 046	104 194	56.32
上 海 Shanghai	281	63 159	95 921	56.77
江 苏 Jiangsu	732	91 284	150 873	61.05
浙 江 Zhejiang	783	103 078	269 609	59.45
福 建 Fujian	390	54 173	88 876	60.79
山 东 Shandong	796	94 734	165 628	63.54
广 东 Guangdong	927	139 488	225 488	58.48
广 西 Guangxi	339	44 805	78 944	58.56
海 南 Hainan	155	30 035	52 491	60.19

注：本表数据为主要沿海城市合计数。
Note: The data in this table are for the total number of major coastal cities.

4-11 沿海地区旅行社数
Number of Travel Agencies by Coastal Regions

单位：家 (unit)

地 区 Region		旅行社总数 Number of Travel Agencies
合 计	**Total**	**12 773**
天 津	Tianjin	344
河 北	Hebei	1 252
辽 宁	Liaoning	1 141
上 海	Shanghai	1 090
江 苏	Jiangsu	1 996
浙 江	Zhejiang	1 894
福 建	Fujian	760
山 东	Shandong	1 963
广 东	Guangdong	1 512
广 西	Guangxi	510
海 南	Hainan	311

主要统计指标解释

1. 海水养殖面积 是指利用海上、滩涂、陆基进行鱼、甲壳类（虾、蟹）、贝、藻等海水经济动植物的人工养殖的水面面积。在报告期内无论是否全部收获或尚未收获其产品，均应统计在海水养殖面积中。但有些滩涂、水面不投放苗种或投放少量苗种，只进行一般管理的，不统计为养殖面积。

2. 盐田总面积 指盐田占有的全部面积。包括储卤、蒸发、保卤、结晶面积、滩内的沟、壕、池、埝、滩坨等面积及滩外的沟、壕、公路及杂地面积。

3. 生产面积 指直接提供给海盐生产的面积，包括结晶面积、蒸发面积、保卤面积，滩内的沟、壕、池、埝面积及滩坨面积。

4. 年末海盐生产能力 指年末企业生产原盐的全部设备的综合平衡能力。海盐生产露天作业，受天气影响，因而计算生产能力时，成熟滩田按 10 年实际平均单位生产面积产量乘以本年成熟滩田生产面积而得，新滩田按设计能力及滩田成熟程度可能达到的产量计算。

5. 海洋渔业船舶 是指配置机器作为动力的从事海洋渔业生产和辅助渔业生产的船舶。

6. 远洋渔船 按我国远洋渔业项目管理办法在非我国管辖海域（外国专属经济区水域或公海）进行常年或季节性生产的渔船。

7. 泊位个数 是指设有系靠船舶设施，在同一时间内可供靠泊最大吨级船舶的艘数。即可靠泊一艘船舶，则计为一个泊位，余类推。泊位分码头泊位和浮筒泊位。

8. 客房数 指饭店实际可用于接待旅游者的房间数。

9. 床位数 指饭店实际可用于接待旅游者的床位数。

Explanatory Notes on Main Statistical Indicators

1. Mariculture Area refers to the area of the water surface where seawater economic animals and plants, such as crustacean (shrimp, crab), shellfish and algae, are cultivated at sea, on tidal flat and land. Whether or not all the products in the area have been harvested or the products have not been harvested yet in the period covered by the report, the area is included in the Mariculture Area. But some tidal flats and water surfaces where none or a small amount of the young have been released and only general management is carried out are not included in the Mariculture Area.

2. Total Area of Salt Pans refers to the total area covered by salt pans, including the area for brine storage, evaporation, brine preservation, and crystallization, the area of ditches, moats, pondsand banks within the beach as well as beach mounds, and ditches, moats, highway beyond the beach as well as the area of miscellaneous lands.

3. Area of Salt Pan Production refers to the area directly provided for sea-salt productions, including the area for crystallization, evaporation and brine preservation, the area of ditches, moats, ponds and banks within the beach as well as the area of beach mounds.

4. Year-End Capacity of Crude Salt Production refers to the integrated and balanced capacity of all equipment of the enterprise used for crude salt production at the end of the year. As sea salt production is an open-air operation, which is subject to the effect of weather, the production capacity of a matured salt pan is calculated at the productions of the actual average unit production area in ten years times the production area of the matured salt pan in the current year. The production capacity of new salt pans is calculated at the production that may be reached in the light of the designed capacity and the level of maturity of the salt pan.

5. Marine Fishing Vessels refer to the vessels equipped with machines as motive power and going for marine fishery production and auxiliary fishery production.

6. Deep Sea Fishing Vessels refer to the fishing vessels which carry out production all the year round or seasonally in the non-Chinese jurisdictional, sea areas (foreign EEZ or high sea) according to the China Deep-Sea Fishing Projects Management Measures.

7. Number of Berths refers to the spaces equipped with facilities for docking ships and the number of ships of the maximum tonnage that may dock or anchor in them. A space for a ship to dock is counted as one berth and the rest are reasoned out by analogy. Berths are divided into wharf berths and buoy berths.

8. Number of Rooms refers to the number of guest rooms actually used by the hotels receiving tourists.

9. Number of Beds refers to the number of beds actually used by the hotels receiving tourists.

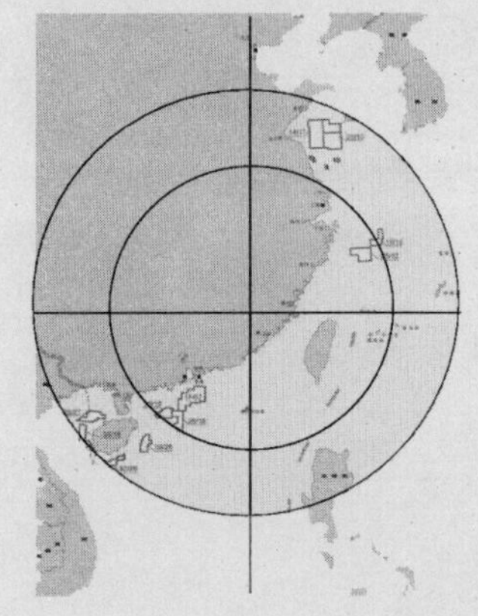

5

涉 海 就 业

Ocean-Related Employment

5-1　全国涉海就业人员情况
National Ocean-Related Employed Personnel

单位：万人　　　　(10 000 persons)

地　区 Region	2001	2011[①]	2012[①]
合　计 Total	**2 107.6**	**3 421.7**	**3 468.8**
天　津 Tianjin	106.4	172.7	175.1
河　北 Hebei	58.0	94.2	95.5
辽　宁 Liaoning	196.0	318.2	322.6
上　海 Shanghai	127.5	207.0	209.8
江　苏 Jiangsu	116.9	189.8	192.4
浙　江 Zhejiang	256.4	416.3	422.0
福　建 Fujian	259.7	421.6	427.4
山　东 Shandong	319.9	519.4	526.5
广　东 Guangdong	505.3	820.4	831.6
广　西 Guangxi	68.9	111.9	113.4
海　南 Hainan	80.6	130.9	132.7
其　他[②] Others	12.0	19.5	19.7

注：① 2011年和2012年为推算数据；② 其他为非沿海地区涉海就业人员数。

Note: ① The data for 2011 and 2012 are the estimated ones;

② Others refer to the number of ocean-related employed persons in the non-coastal regions.

5-2 全国主要海洋产业就业人员情况
Employed Personnel in the Major Marine Industries Throughout the Country

单位：万人 (10 000 persons)

海洋产业 Marine Industry	2001	2011	2012
合 计 Total	**719.1**	**1 167.5**	**1 183.5**
海洋渔业及相关产业 Marine Fishery and the Related Industries	348.3	565.5	573.2
海洋石油和天然气业 Offshore Oil and Natural Gas Industry	12.4	20.1	20.4
海滨砂矿业 Beach Placer Mining Industry	1.0	1.6	1.6
海洋盐业 Sea Salt Industry	15.0	24.4	24.7
海洋化工业 Marine Chemical Industry	16.1	26.1	26.5
海洋生物医药业 Marine Biomedicine Industry	0.6	1.0	1.0
海洋电力和海水利用业 Marine Electric Power and Seawater Utilization Industry	0.7	1.1	1.2
海洋船舶工业 Marine Shipbuilding Industry	20.6	33.4	33.9
海洋工程建筑业 Marine Engineering Architecture Industry	38.8	63.0	63.9
海洋交通运输业 Maritime Communications and Transportation Industry	50.8	82.5	83.6
滨海旅游业 Coastal Tourism	78.3	127.1	128.9
其他海洋产业 Other Marine Industries	136.5	221.6	224.7

注：2011年和2012年为推算数据。

Note: The data for 2011 and 2012 are the estimated ones.

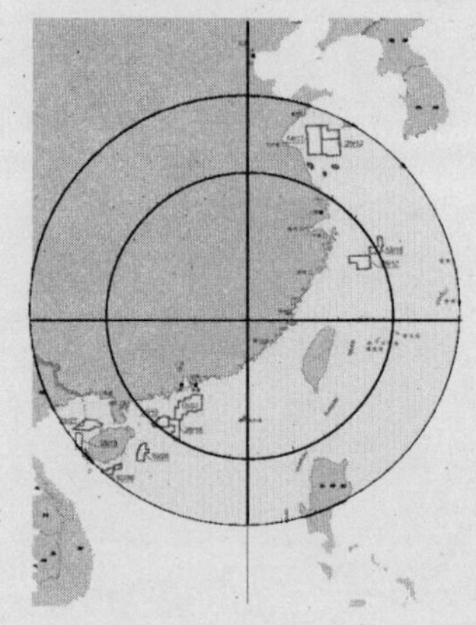

6

海洋科学技术

Marine Science and Technology

6-1 分行业海洋科研机构及人员情况
Marine Scientific Research Institutions and Personnel by Industry

行 业 Industry	机构数（个） Number of Institutions (unit)	从业人员（人） Employed Population (person)
合 计 **Total**	**177**	**37 679**
海洋基础科学研究 **Marine Basic Scientific Research**	**100**	**17 508**
海洋自然科学 Marine Natural Science	56	13 280
海洋社会科学 Marine Social Science	4	812
海洋农业科学 Marine Agricultural Science	38	3 349
海洋生物医药 Marine Biomedicine	2	67
海洋工程技术研究 **Marine Engineering Technology Research**	**65**	**18 495**
海洋化学工程技术 Marine Chemical Engineeing Technology	12	6 326
海洋生物工程技术 Marine Bioengineering Technology	2	257
海洋交通运输工程技术 Marine Communications and Transport Technology	14	3 659

6-1 续表 continued

行 业 Industry	机构数（个） Number of Institutions (unit)	从业人员（人） Employed Population (person)
海洋能源开发技术 Marine Energies Development Technology	4	3 177
海洋环境工程技术 Marine Environmental Engineering Technology	11	1 030
河口水利工程技术 Eustuarine Water Conservancy Engineering Technology	18	3 214
其他海洋工程技术 Other Marine Engineering Technology	4	832
海洋信息服务业 **Marine Information Service**	**9**	**1 017**
其他海洋信息服务 Other Marine Information Services	9	1 017
海洋技术服务业 **Marine Technological Service Industry**	**3**	**659**
其他海洋专业技术服务 Other Marine Professional and Technological Services	2	123
海洋工程管理服务 Marine Engineering Management Service	1	536

注：机构为县级以上科研机构；行业分类参照《海洋及相关产业分类》标准。

Note: The institutions are the scientific research institutions above the county level;The classification of industries follows the standard *Classification of Marine Industries and the Related Industries* .

6-2 分行业海洋科研机构科技活动人员学历构成
Educational Background Composition of the Personnel Engaged in Scientific and Technological Activities in Marine Scientific Research Institutions by Industry

单位：人 (person)

行 业 Industry	科技活动人员 Personnel Engaged in Scientific Activities	博士 Doctor	硕士 Master	大学生 University Graduate	大专生 College Graduate
合 计 Total	**31 487**	**6 983**	**8 819**	**9 943**	**3 575**
海洋基础科学研究 Marine Basic Scientific Research	**15 370**	**4 591**	**3 978**	**4 145**	**1 445**
海洋自然科学 Marine Natural Science	11 808	4 123	3 029	2 809	1 033
海洋社会科学 Marine Social Science	774	106	280	288	78
海洋农业科学 Marine Agricultural Science	2 739	361	664	1 017	328
海洋生物医药 Marine Biomedicine	49	1	5	31	6
海洋工程技术研究 Marine Engineering Technology Research	**14 593**	**2 285**	**4 292**	**5 127**	**1 986**
海洋化学工程技术 Marine Chemical Engineeing Technology	5 327	730	1 213	1 694	1 139
海洋生物工程技术 Marine Bioengineering Technology	200	12	85	73	22

6-2 续表 continued

行 业 Industry	科技活动人员 Personnel Engaged in Scientific Activities	博士 Doctor	硕士 Master	大学生 University Graduate	大专生 College Graduate
海洋交通运输工程技术 Marine Communications and Transport Technology	2 273	112	699	1 106	226
海洋能源开发技术 Marine Energies Development Technology	2 819	871	947	716	240
海洋环境工程技术 Marine Environmental Engineering Technology	837	52	239	445	71
河口水利工程技术 Eustuarine Water Conservancy Engineering Technology	2 588	438	848	928	254
其他海洋工程技术 Other Marine Engineering Technology	549	70	261	165	34
海洋信息服务业 Marine Information Service	**878**	**72**	**309**	**347**	**107**
其他海洋信息服务 Other Marine Information Services	878	72	309	347	107
海洋技术服务业 Marine Technological Service Industry	**646**	**35**	**240**	**324**	**37**
其他海洋专业技术服务 Other Marine Professional and Technological Services	110	4	14	69	16
海洋工程管理服务 Marine Engineering Management Service	536	31	226	255	21

6-3 分行业海洋科研机构科技活动人员职称构成
Professional Title Composition of Personnel Engaged in Scientific and Technological Activities in the Marine Scientific Research Institutions by Industry

单位：人 (person)

行业 Industry	科技活动人员 Personnel Engaged in Scientific Activities	高级职称 Senior	中级职称 Intermediate	初级职称 Primary
合计 Total	**31 487**	**12 360**	**10 677**	**5 130**
海洋基础科学研究 Marine Basic Scientific Research	**15 370**	**6 092**	**5 360**	**2 580**
海洋自然科学 Marine Natural Science	11 808	4 886	4 173	1 916
海洋社会科学 Marine Social Science	774	328	220	60
海洋农业科学 Marine Agricultural Science	2 739	873	952	575
海洋生物医药 Marine Biomedicine	49	5	15	29
海洋工程技术研究 Marine Engineering Technology Research	**14 593**	**5 720**	**4 736**	**2 240**
海洋化学工程技术 Marine Chemical Engineeing Technology	5 327	1 946	1 676	666
海洋生物工程技术 Marine Bioengineering Technology	200	50	64	74

6-3 续表 continued

行 业 Industry	科技活动人员 Personnel Engaged in Scientific Activities	高级职称 Senior	中级职称 Intermediate	初级职称 Primary
海洋交通运输工程技术 Marine Communications and Transport Technology	2 273	642	731	632
海洋能源开发技术 Marine Energies Development Technology	2 819	1 457	827	302
海洋环境工程技术 Marine Environmental Engineering Technology	837	285	351	155
河口水利工程技术 Eustuarine Water Conservancy Engineering Technology	2 588	1 131	848	334
其他海洋工程技术 Other Marine Engineering Technology	549	209	239	77
海洋信息服务业 Marine Information Service	**878**	**323**	**292**	**207**
其他海洋信息服务 Other Marine Information Services	878	323	292	207
海洋技术服务业 Marine Technological Service Industry	**646**	**225**	**289**	**103**
其他海洋专业技术服务 Other Marine Professional and Technological Services	110	54	36	15
海洋工程管理服务 Marine Engineering Management Service	536	171	253	88

6-4 分行业海洋科研机构经费收入
Research Funds Receipts of the Marine Scientific Research Institutions by Industry

单位：万元 (10 000 yuan)

行 业 Industry	经费收入总额 Fund Total	经常费 Routine Fund	科技活动借贷款 Loans in the Scientific and Technological Activities	基本建设中政府投资 Government Investment in the Capital Construction
合 计 **Total**	**25 772 307**	**23 853 523**	**252 064**	**1 666 720**
海洋基础科学研究 **Marine Basic Scientific Research**	**12 603 751**	**11 410 388**	**164**	**1 193 199**
海洋自然科学 Marine Natural Science	10 845 342	9 745 397	164	1 099 781
海洋社会科学 Marine Social Science	446 620	401 396	0	45 224
海洋农业科学 Marine Agricultural Science	1 296 540	1 248 346	0	48 194
海洋生物医药 Marine Biomedicine	15 249	15 249	0	0
海洋工程技术研究 **Marine Technological Service Industry**	**12 436 442**	**11 737 379**	**251 900**	**447 163**
海洋化学工程技术 Marine Chemical Engineeing Technology	4 010 403	3 801 888	46 000	162 515
海洋生物工程技术 Marine Bioengineering Technology	164 510	164 510	0	0

6-4 续表 continued

行 业 Industry	经费收入总额 Fund Total	经常费 Routine Fund	科技活动借贷款 Loans in the Scientific and Technological Activities	基本建设中政府投资 Government Investment in the Capital Construction
海洋交通运输工程技术 Marine Communications and Transport Technology	2 718 550	2 286 573	205 900	226 077
海洋能源开发技术 Marine Energies Development Technology	2 924 740	2 924 740	0	0
海洋环境工程技术 Marine Environmental Engineering Technology	497 776	495 657	0	2 119
河口水利工程技术 Eustuarine Water Conservancy Engineering Technology	1 628 739	1 572 287	0	56 452
其他海洋工程技术 Other Marine Engineering Technology	491 724	491 724	0	0
海洋信息服务业 **Marine Information Service**	**458 593**	**432 235**	**0**	**26 358**
其他海洋信息服务 Other Marine Information Services	458 593	432 235	0	26 358
海洋技术服务业 **Marine Technological Service Industry**	**273 521**	**273 521**	**0**	**0**
其他海洋专业技术服务 Other Marine Professional and Technological Services	200 622	200 622	0	0
海洋工程管理服务 Marine Engineering Management Service	72 899	72 899	0	0

6-5 分行业海洋科研机构科技课题情况

Marine Scientific and Technological Research Projects of the Research Institutions by Industry

单位：项 (item)

行业 Industry	课题数 Number of Research Projects	基础研究 Basic Research	应用研究 Applied Research	试验发展 Experimental Development	成果应用 Result Application	科技服务 Scientific and Technological Service
合计 Total	**15 403**	**3 756**	**3 854**	**3 569**	**1 515**	**2 709**
海洋基础科学研究 Marine Basic Scientific Research	**11 421**	**3 694**	**3 323**	**2 034**	**899**	**1 471**
海洋自然科学 Marine Natural Science	9 198	3 562	2 861	1 376	517	882
海洋社会科学 Marine Social Science	613	8	21	133	108	343
海洋农业科学 Marine Agricultural Science	1 595	123	438	519	269	246
海洋生物医药 Marine Biomedicine	15	1	3	6	5	0
海洋工程技术研究 Marine Engineering Technology Research	**3 719**	**62**	**527**	**1 497**	**610**	**1 023**
海洋化学工程技术 Marine Chemical Engineeing Technology	962	18	153	573	156	62
海洋生物工程技术 Marine Bioengineering Technology	8	0	0	0	1	7

6-5 续表 continued

行 业 Industry	课题数 Number of Research Projects	基础研究 Basic Research	应用研究 Applied Research	试验发展 Experimental Development	成果应用 Result Application	科技服务 Scientific and Technological Service
海洋交通运输工程技术 Marine Communications and Transport Technology	711	5	44	224	196	242
海洋能源开发技术 Marine Energies Development Technology	695	8	176	276	50	185
海洋环境工程技术 Marine Environmental Engineering Technology	223	1	30	56	57	79
河口水利工程技术 Eustuarine Water Conservancy Engineering	1 019	28	118	322	132	419
其他海洋工程技术 Other Marine Engineering Technology	101	2	6	46	18	29
海洋信息服务业 Marine Information Service	**172**	**0**	**2**	**29**	**2**	**139**
其他海洋信息服务 Other Marine Information Services	172	0	2	29	2	139
海洋技术服务业 Marine Technological Service Industry	**91**	**0**	**2**	**9**	**4**	**76**
其他海洋专业技术服务 Other Marine Professional and Technological Services	15	0	2	9	4	0
海洋工程管理服务 Marine Engineering Management Service	76	0	0	0	0	76

6-6 分行业海洋科研机构科技论著情况

Marine Scientific and Technological Works of the Research Institutions by Industry

行　业 Industry	发表科技论文（篇） Scientific Theses Published (piece)	#国外发表 Published Abroad	出版科技著作（种） Scientific and Technological Works Published (kind)
合　计 Total	**16 713**	**4 672**	**338**
海洋基础科学研究 Marine Basic Scientific Research	**11 194**	**3 949**	**207**
海洋自然科学 Marine Natural Science	8 812	3 696	123
海洋社会科学 Marine Social Science	632	28	55
海洋农业科学 Marine Agricultural Science	1 740	221	29
海洋生物医药 Marine Biomedicine	10	4	0
海洋工程技术研究 Marine Engineering Technology Research	**5 135**	**708**	**121**
海洋化学工程技术 Marine Chemical Engineeing Technology	1 005	131	4
海洋生物工程技术 Marine Bioengineering Technology	89	5	2

6-6 续表 continued

行 业 Industry	发表科技论文（篇） Scientific Theses Published (piece)	#国外发表 Published Abroad	出版科技著作（种） Scientific and Technological Works Published (kind)
海洋交通运输工程技术 Marine Communications and Transport Technology	741	93	9
海洋能源开发技术 Marine Energies Development Technology	1 112	219	24
海洋环境工程技术 Marine Environmental Engineering Technology	234	13	7
河口水利工程技术 Eustuarine Water Conservancy Engineering	1 525	234	68
其他海洋工程技术 Other Marine Engineering Technology	429	13	7
海洋信息服务业 **Marine Information Service**	**289**	**13**	**3**
其他海洋信息服务 Other Marine Information Services	289	13	3
海洋技术服务业 **Marine Technological Service Industry**	**95**	**2**	**7**
其他海洋专业技术服务 Other Marine Professional and Technological Services	18	2	0
海洋工程管理服务 Marine Engineering Management Service	77	0	7

6-7 分行业科研机构科技专利情况
Marine Scientific and Technological Patents of the Research Institutions by Industry

单位：件 (unit)

行 业 Industry	专利申请受理数 Number of Patent Applications Accepted	#发明专利 Patents for Discoveries	专利授权数 Number of Patents Granted	#发明专利 Patents for Discoveries	拥有发明专利总数 Total Number of Patents for Discoveries
合 计 Total	**5 120**	**4 202**	**2 746**	**1 883**	**10 695**
海洋基础科学研究 Marine Basic Scientific Research	**1 690**	**1 262**	**1 267**	**831**	**3 455**
海洋自然科学 Marine Natural Science	1 260	999	952	669	3 128
海洋社会科学 Marine Social Science	2	0	0	0	6
海洋农业科学 Marine Agricultural Science	428	263	315	162	321
海洋生物医药 Marine Biomedicine	0	0	0	0	0
海洋工程技术研究 Marine Engineering Technology Research	**3 413**	**2 925**	**1 457**	**1 036**	**7 201**
海洋化学工程技术 Marine Chemical Engineeing Technology	2 655	2 520	999	869	6 105
海洋生物工程技术 Marine Bioengineering Technology	2	2	0	0	0

6-7 续表 continued

行 业 Industry	专利申请受理数 Number of Patent Applications Accepted	#发明专利 Patents for Discoveries	专利授权数 Number of Patents Granted	#发明专利 Patents for Discoveries	拥有发明专利总数 Total Number of Patents for Discoveries
海洋交通运输工程技术 Marine Communications and Transport Technology	84	40	63	24	125
海洋能源开发技术 Marine Energies Development Technology	354	206	162	67	469
海洋环境工程技术 Marine Environmental Engineering Technology	23	8	24	9	16
河口水利工程技术 Eustuarine Water Conservancy Engineering Technology	125	48	126	48	281
其他海洋工程技术 Other Marine Engineering Technology	170	101	83	19	205
海洋信息服务业 Marine Information Service	**4**	**4**	**1**	**1**	**1**
其他海洋信息服务 Other Marine Information Services	4	4	1	1	1
海洋技术服务业 Marine Technological Service Industry	**13**	**11**	**21**	**15**	**38**
其他海洋专业技术服务 Other Marine Professional and Technological Services	13	11	19	15	37
海洋工程管理服务 Marine Engineering Management Service	0	0	2	0	1

6-8 分行业海洋科研机构R & D情况
R & D in the Marine Scientific Research Institutions by Profession

行　业 Industry	R & D人员 （人） R & D Personnel (persons)	R & D经费内部支出 （千元） R & D Internal Expenditure (1 000 yuan)	R & D课题数 （项） Number of R & D Projects (item)
合　计 **Total**	**26 151**	**12 264 655**	**11 179**
海洋基础科学研究 **Marine Basic Scientific Research**	**17 994**	**7 741 450**	**9 051**
海洋自然科学 Marine Natural Science	15 584	7 077 583	7 799
海洋社会科学 Marine Social Science	313	80 211	162
海洋农业科学 Marine Agricultural Science	2 070	578 910	1 080
海洋生物医药 Marine Biomedicine	27	4 746	10
海洋工程技术研究 **Marine Engineering Technology Research**	**7 717**	**4 351 474**	**2 086**
海洋化学工程技术 Marine Chemical Engineeing Technology	3 237	1 979 978	744
海洋生物工程技术 Marine Bioengineering Technology	0	0	0

6-8 续表 continued

行　业 Industry	R & D人员 （人） R & D Personnel (persons)	R & D经费内部支出 （千元） R & D Internal Expenditure (1 000 yuan)	R & D课题数 （项） Number of R & D Projects (item)
海洋交通运输工程技术 Marine Communications and Transport Technology	1 182	315 184	273
海洋能源开发技术 Marine Energies Development Technology	1 728	1 422 370	460
海洋环境工程技术 Marine Environmental Engineering Technology	263	104 853	87
河口水利工程技术 Eustuarine Water Conservancy Engineering Technology	1 022	416 278	468
其他海洋工程技术 Other Marine Engineering Technology	285	112 811	54
海洋信息服务业 Marine Information Service	**368**	**152 692**	**31**
其他海洋信息服务 Other Marine Information Services	368	152 692	31
海洋技术服务业 Marine Technological Service Industry	**72**	**19 039**	**11**
其他海洋专业技术服务 Other Marine Professional and Technological Services	72	19 039	11
海洋工程管理服务 Marine Engineering Management Service	0	0	0

6-9　分地区海洋科研机构及人员情况
Marine Scientific Research Institutions and Personnel by Regions

地　区 Region	机构数（个） Number of Institutions (unit)	从业人员（人） Employed Population (person)
合 计 Total	**177**	**37 679**
北 京 Beijing	25	13 857
天 津 Tianjin	14	2 628
河 北 Hebei	5	552
辽 宁 Liaoning	17	2 077
上 海 Shanghai	14	3 721
江 苏 Jiangsu	11	2 900
浙 江 Zhejiang	18	1 695
福 建 Fujian	12	1 075
山 东 Shandong	21	3 818
广 东 Guangdong	24	3 164
广 西 Guangxi	9	444
海 南 Hainan	3	192
其 他 Other	4	1 556

6-10 分地区海洋科研机构科技活动人员学历构成

Educational Background Composition of the Personnel Engaged in Scientific and Technological Activities in Marine Scientific Research Institutions by Regions

单位：人 (person)

地 区 Region	科技活动人员 Personnel Engaged in Scientifical Activities	博士 Doctor	硕士 Master	大学生 University Graduate	大专生 College Graduate
合 计 Total	**31 487**	**6 983**	**8 819**	**9 943**	**3 575**
北 京 Beijing	12 349	3 604	3 551	3 226	1 431
天 津 Tianjin	2 116	155	642	936	201
河 北 Hebei	531	36	127	280	57
辽 宁 Liaoning	1 662	134	411	688	243
上 海 Shanghai	3 127	504	953	1 089	362
江 苏 Jiangsu	1 762	308	480	622	166
浙 江 Zhejiang	1 407	133	468	610	150
福 建 Fujian	1 015	114	292	385	127
山 东 Shandong	3 203	741	845	921	395
广 东 Guangdong	2 638	724	655	684	297
广 西 Guangxi	358	13	68	211	60
海 南 Hainan	179	6	39	55	19
其 他 Other	1 140	511	288	236	67

6-11 分地区海洋科研机构科技活动人员职称构成
Professional Title Composition of Personel Engaged in Marine Scientific Research Institutions by Regions

单位：人 (person)

地区 Region	科技活动人员 Personnel Engaged in Scientifical Activities	高级职称 Senior	中级职称 Intermediate	初级职称 Primary
合计 Total	**31 487**	**12 360**	**10 677**	**5 130**
北京 Beijing	12 349	5 629	4 131	1 440
天津 Tianjin	2 116	718	764	436
河北 Hebei	531	200	128	36
辽宁 Liaoning	1 662	563	528	182
上海 Shanghai	3 127	1 054	1 112	711
江苏 Jiangsu	1 762	720	503	334
浙江 Zhejiang	1 407	513	493	266
福建 Fujian	1 015	303	373	248
山东 Shandong	3 203	1 113	1 125	691
广东 Guangdong	2 638	957	877	451
广西 Guangxi	358	73	141	116
海南 Hainan	179	17	25	60
其他 Other	1 140	500	477	159

6-12 分地区海洋科研机构经费收入
Research Funds Receipts of Marine Scientific Research Institutions by Regions

单位：万元 (10 000 yuan)

地 区 Region	经费收入总额 Fund Total	经常费 Routine Fund	科技活动借贷款 Loans in the Scientific and Technological Activities	基本建设中政府投资 Government Investment in the Capital Construction
合 计 Total	**25 772 307**	**23 853 523**	**252 064**	**1 666 720**
北 京 Beijing	9 706 492	9 349 443	0	357 049
天 津 Tianjin	1 636 931	1 415 593	0	221 338
河 北 Hebei	124 021	122 409	0	1 612
辽 宁 Liaoning	1 134 138	1 115 788	0	18 350
上 海 Shanghai	2 897 953	2 760 743	0	137 210
江 苏 Jiangsu	2 009 956	1 783 037	200 000	26 919
浙 江 Zhejiang	1 318 163	1 284 446	5 900	27 817
福 建 Fujian	846 320	844 825	0	1 495
山 东 Shandong	3 166 172	2 559 462	38 164	568 546
广 东 Guangdong	1 774 881	1 632 533	8 000	134 348
广 西 Guangxi	93 599	93 599	0	0
海 南 Hainan	99 504	76 981	0	22 523
其 他 Other	964 177	814 664	0	149 513

6-13 分地区海洋科研机构科技课题情况
Marine Scientific and Technological Research Projects of the Research Institutions by Regions

单位：项 (item)

地 区 Region	课题数 Number of Research Projects	基础研究 Basic Research	应用研究 Applied Research	试验发展 Experimental Development	成果应用 Result Application	科技服务 Scientific and Technological Service
合 计 Total	**15 403**	**3 756**	**3 854**	**3 569**	**1 515**	**2 709**
北 京 Beijing	5 412	1 475	1 271	1 152	331	1 183
天 津 Tianjin	668	14	111	265	82	196
河 北 Hebei	78	8	23	20	11	16
辽 宁 Liaoning	337	14	43	148	91	41
上 海 Shanghai	996	44	272	333	128	219
江 苏 Jiangsu	1 851	121	406	698	462	164
浙 江 Zhejiang	494	53	94	102	104	141
福 建 Fujian	591	160	238	29	74	90
山 东 Shandong	1 550	347	627	350	95	131
广 东 Guangdong	2 190	722	601	310	67	490
广 西 Guangxi	106	5	20	69	8	4
海 南 Hainan	46	0	0	2	43	1
其 他 Other	1 084	793	148	91	19	33

6-14 分地区海洋科研机构科技论著情况
Marine Scientific and Technological Works of the Research Institutions by Regions

地 区 Region	发表科技论文（篇） Scientific Theses Published (piece)	#国外发表 Published Abroad	出版科技著作（种） Scientific and Technological Works Published (kind)
合 计 Total	**16 713**	**4 672**	**338**
北 京 Beijing	6 271	1 930	138
天 津 Tianjin	851	113	15
河 北 Hebei	448	4	40
辽 宁 Liaoning	478	40	8
上 海 Shanghai	1 223	228	14
江 苏 Jiangsu	1 040	260	16
浙 江 Zhejiang	509	77	14
福 建 Fujian	350	63	13
山 东 Shandong	2 023	674	38
广 东 Guangdong	2 104	734	22
广 西 Guangxi	105	7	0
海 南 Hainan	69	5	3
其 他 Other	1 242	537	17

6-15 分地区海洋科研机构科技专利情况
Marine Scientific and Technological Patents of the Research Institutions by Regions

单位：件 (unit)

地 区 Region	专利申请受理数 Number of Patent Applications Accepted	#发明专利 Patents for Discoveries	专利授权数 Number of Patents Granted	#发明专利 Patents for Discoveries	拥有发明专利总数 Total Number of Patents for Discoveries
合 计 Total	**5 120**	**4 202**	**2 746**	**1 883**	**10 695**
北 京 Beijing	2 460	2 160	1 135	906	5 378
天 津 Tianjin	130	59	79	36	145
河 北 Hebei	8	4	6	2	4
辽 宁 Liaoning	616	552	252	181	1 268
上 海 Shanghai	842	687	435	295	1 472
江 苏 Jiangsu	125	75	105	43	204
浙 江 Zhejiang	47	35	59	33	116
福 建 Fujian	24	11	16	10	121
山 东 Shandong	401	304	280	165	553
广 东 Guangdong	273	210	246	165	1 065
广 西 Guangxi	16	15	13	10	21
海 南 Hainan	2	1	5	4	5
其 他 Other	176	89	115	33	343

6-16 分地区海洋科研机构R & D情况
R & D in the Marine Scientific Research Institutions by Regions

行 业 Industry	R & D人员 （人） R & D Personnel (persons)	R & D经费内部支出 （千元） R & D Internal Expenditure (1 000 yuan)	R & D课题数 （项） Number of R & D Projects (item)
合 计 **Total**	**26 151**	**12 264 655**	**11 179**
北 京 Beijing	9 985	5 154 263	3 898
天 津 Tianjin	1 501	652 666	390
河 北 Hebei	251	37 411	51
辽 宁 Liaoning	865	396 628	205
上 海 Shanghai	2 578	1 434 100	649
江 苏 Jiangsu	1 597	494 402	1 225
浙 江 Zhejiang	610	337 108	249
福 建 Fujian	478	255 753	427
山 东 Shandong	2 879	1 728 892	1 324
广 东 Guangdong	3 167	1 015 391	1 633
广 西 Guangxi	202	40 617	94
海 南 Hainan	4	110	2
其 他 Other	2 034	717 314	1 032

主要统计指标解释

1. 海洋科研机构 指有明确的研究方向和任务，有一定水平的学术带头人和一定数量、质量的研究人员，有开展研究工作的基本条件，长期有组织地从事海洋研究与开发活动的机构。

2. 从业人员 指由本机构年末直接组织安排工作并支付工资的各类人员总数。包括固定职工、国家有编制的合同制职工、招聘人员和返聘的离退休人员。不包括离退休人员、停薪留职人员。

3. 从事科技活动人员 指从业人员中的科技管理人员、课题活动人员和科技服务人员。

4. 高级职称 指研究员、副研究员；教授、副教授；高级工程师；高级农艺师；正、副主任医（药、护、技)师；高级实验师；高级统计师；高级经济师；高级会计师；编审(正、副编审)；译审(正、副译审)、高级(主任)记者；正、副研究馆员等。

5. 中级职称 指助理研究员；讲师；工程师；农艺师；主治医(药、护、技)师；实验师；统计师；经济师；会计师；编辑；翻译；记者；馆员等。

6. 初级职称 指研究实习员；助教；助理工程师、技术员；助理农艺师、农业技术员；医(药、护、技)师、医(药、护、技)士；助理实验师、实验员；助理统计师、统计员；助理经济师；助理会计师、会计员；助理编辑、见习编辑；助理翻译；助理记者；助理馆员、管理员等。

7. 科技经费筹集额 指从各种渠道筹集到的计划用于本单位科技活动的经费，不论来源渠道如何。

8. 政府资金 指由各级政府部门直接拨款或企事业单位利用政府资金委托本机构从事科学技术活动所获得的收入。

9. 生产经营活动收入 指本机构在科研、技术等专业业务活动以外开展非独立核算的经营活动取得的收入，包括产品（商品）销售收入、经营服务收入、工程承包收入、租赁收入和其他经营收入。

10. 其他收入 指开展科技活动与生产经营活动以外的各项活动的收入，包括：用于离退休人员的政府拨款。

11. 非科技活动借贷款 指本机构为开展非科技活动从各种渠道获得的各类借、贷款。不论偿还形式、期限和数额如何，均按当年获得的借、贷款额填报。不包括基本建设贷款。

12. 基础研究 为获得新知识而进行的独创性研究。其目的是揭示观察到的现象和事实的基本原理和规律，而不以任何特定的实际应用为目的。

13. 应用研究 为获得新的科学技术知识而进行的独创性研究。它主要针对某一特定的实际应用目的。应用研究通常是为了确定基础研究成果或知识的可能的用途，或是为达到某一具体的、预定的实际目的确定新的方法(原理性)或途径。

14. 试验发展 利用从研究或实际经验获得的知识，为生产新的材料、产品和装置，建立新的工艺和系统，以及对已生产或建立的上述各项进行实质性的改进而进行的系统性工作。

15. 成果应用 为解决 R&D 活动阶段产生的新产品、新装置、新工艺、新技术、新方法、新系统和服务等能投入生产或在实际应用中所存在的技术问题而进行的系统性活动。它不具有创新成分。此类活动包括为达到生产目的而进行的定型设计和试制以及为扩大新产品的生产规模和探索新方法、新技术、新工艺等的应用领域而进行的适应性试验。

16. 科技服务 与科学研究与实验发展有关，并有助于科学技术知识的产生、传播和应用的活动。包括为扩大科技成果的使用范围而进行的示范性推广工作；为用户提供科技信息和文献服务的系统性工作；为用户提供可行性报告、技术方案、建议及进行技术论证等技术咨询工作；自然、生物现象的日常观测、监测，资源的考查和勘探；有关社会、人文、经济现象的通用资料的收集，如统计、市场调查等以及这些资料的常规分析与整理；为社会和公众提供的测试、标准化、计量、计算、质量控制和专利服务，不包括工商企业为进行正常生产而开展的上述活动。

17. 生产性活动 由于业务特殊的工艺设备条件，或掌握某种技术专长或诀窍，所进行的小量非常规生产。

18. 科技论文 在全国性学报或学术刊物上、省部属大专院校对外正式发行的学报或学术刊物上发表的论文以及向国外发表的论文。

19. 科技著作 经过正式出版部门编印出版的科技专著、大专院校教科书、科普著作。

20. 专利申请受理数 当年本单位向专利管理部门提出申请并被受理的职务专利申请件数。

21. 专利授权数 当年由专利管理部门授予本单位专利权的职务专利件数。

Explanatory Notes on Main Statistical Indicators

1. Marine Scientific Research Institution refers to the institution which has definite research orientations and tasks, high-level academic leading personnel and fair-sized, qualified research personnel, and basic conditions for research work and which is engaged for a long time in the marine research and development activities in an organized way.

2. Employees refer to the total number of personnel of various kinds employed and paid by the institution at the end of the year, including fixed employees, contract workers of staff belonging to the state authorized staff , recruited personnel, reemployed retired personnel, but not including the retired and the personnel on leave with pay suspension.

3. Personnel Engaged in Scientific and Technological Activities refers to the personnel for scientific and Technological management, personnel engaged in the activities of research topics and scientific and technological service personnel.

4. Senior Technical Title refer to research scientist, associate research scientist; professor, associate professor; senior engineer; senior agronomist; professor-rank and associate professor-rank doctor (pharmacists, nurses and technicians); senior laboratory technician; senior statisticians; senior economic engineer; chief accountant; senior editor (professor and associate professor ranks); senior translator (professor and associate professor ranks); senior journalist; research librarian (professor and associate professor ranks), etc.

5. Intermediate Technical Title refer to assistant research scientist; lecturer; engineer; agronomist; lecturer-rank doctor (pharmacist, nurse and technician); laboratory technician; lecturer-rank statistician;

economic engineer; accountant; editor; translator; journalist; librarian, etc.

6. Primary Technical Title refers to trainee researcher; assistant; assistant engineer, technician, assistant agronomist, agricultural technician; assistant-rank doctor (pharmacist, nurse and technician); assistant laboratory technician; assistant statistician; assistant economic engineer, assistant accountant; assistant editor, editor on probation; assistant translator; assistant journalist; assistant research librarian, librarian ,etc.

7. Amount of Scientific and Technological Funds Raised refers to the funds raised through all channels planned to be used as funds for the scientific and technological activities in the institution regardless of their source and channels.

8. Funds from government refers to the direct appropriations by the government departments at all levels or the earnings from conducting scientific and technological activities entrusted to the institution by enterprises as institutions by earning the funds from government.

9. Earnings from Production as Business Activities refer to the incomes obtained from the non-independent accounting business activities carried out by the institution beyond the scientific research, technological and professional activities, including these from sale of products(goods), business and service, contracted projects, leasing and other business.

10. Other Incomes refer to those from the various activities conducted other than scientific and technological activities and production and business activities, including the government appropriations for retired personnel.

11. Loan for Non-scientific as Technological Activities refers to the various types loan obtained by the institutions through all channels for carrying out non-scientific and technological activities, not including the load for capital construction. The loan, irrespective of its form of reimbursement, term and amount is filled in a form and submitted to the authorities as the amount acquired in the current year.

12. Basic Research refers to the original research to acquire new knowledge. It is aimed at revealing the basic principles and laws of the phenomena and facts observed, but not at any specific practical applications.

13. Applied Research refers to the original research to acquire new scientific and technological knowledge. It mainly serves the purpose of a particular practical application. The purpose of applied research is usually to define the potential uses of the research finds or knowledge obtained from basic research or to identify new methods (principles) or ways to reach a specific and predetermined goal.

14. Experimental Development refers to the systematic work carried out to establish new technologies and systems for producing new materials, products and equipment by using the knowledge obtained from research or practical experience, or to make substantial improvement of the above-mentioned which have been produced or established.

15. Result Application refers to the systematic activities carried out to solve the technical problems that might crop up in the production or practical application of the new products, devices, technologies, techniques, methods, systems and service occurring in the course of R&D activities.

They do not bring forth new ideas. Such activities include the finalizing design and trial-production for the purpose of production as well as the adaptive tests to expand the production scale of new products and the application areas of new methods, techniques and technologies.

16. Scientific and Technological Service refers to the activities that are associated with the scientific research and experimental development, and contribute to the generation, dissemination and application of scientific and technological knowledge, which include the demonstrative work of popularization to enlarge the use scope of scientific and technological achievements; the systematic work of providing the users with scientific and technological information and literature service; the technical consultation work of providing users with feasibility reports, technical schemes and recommendations and carrying out technical demonstration; routine observation and monitoring of natural and biological phenomena, and the survey and exploration of resources; collection of universal data on the appropriate social, cultural and economic phenomena, such as statistics and market survey, as well as the routine analysis and sorting-out of these data; the provision for the society and the public of such service as testing, standardization, computation, quality control and patent, but not including the type of the above-mentioned activities carried out by industrial and commercial enterprises for the purpose of normal production.

17. Productive Activity refers to the small-scale and non-conventional productions due to the presence of special technologies and equipment or mastery of a particular technical expertise or secret of success.

18. Scientific Treatises refer to the theses published in the national journals or academic publications, those officially issued journals or academic publications by universities and colleges under provinces or ministries as well as theses published abroad.

19. Scientific and Technological Works refer to the scientific and technological monographs, textbooks for universities and colleges and popular science books published by the official publishing houses.

20. Number of Patents Applied and Accepted refers to the number of professional patent applications of the unit to the patent administrative department and accepted by it in the year.

21. Number of Patents Granted refers to the number of the professional patents granted to the unit by the patent administrative department in the year.

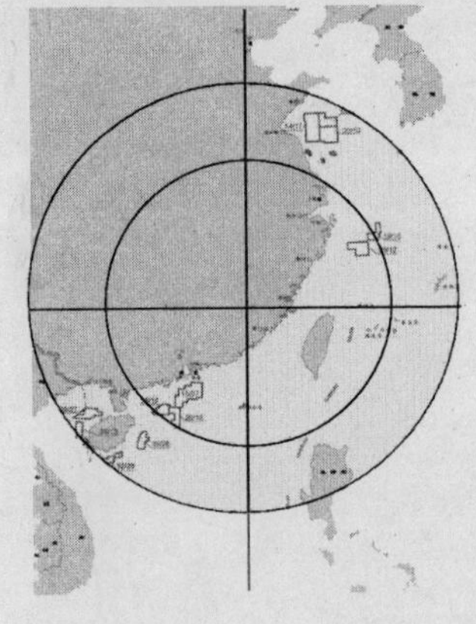

7

海 洋 教 育

Marine Education

7-1 全国各海洋专业博士研究生情况
Doctoral Students from Marine Specialities

专业 Speciality	专业点数（个） Number of Speciality Agencies (unit)	学生数（人） Number of Students (person)			
		毕业生 Graduates	招生 Entrants	在校生 Enrollment	预计毕业生数 Estimated Graduates of Next Year
合计 **Total**	**131**	**615**	**870**	**3 660**	**1 806**
物理海洋学 Physical Oceanography	6	46	61	235	119
海洋化学 Marine Chemistry	5	28	30	130	71
海洋生物学 Marine Biology	9	76	87	284	128
海洋地质 Marine Geology	10	34	54	231	123
海洋科学新专业 New Speciality of Marine Science	4	72	85	304	150
港口海岸及近海工程 Coastal Harbour and Offshore Engineering	10	37	55	303	148
船舶与海洋结构物设计制造 Ships and Marine Structures Design and Manufacture	9	39	84	331	131
轮机工程 Turbine Engineering	8	26	43	259	82

7-1 续表 continued

专 业 Speciality	专业点数（个） Number of Speciality Agencies (unit)	学生数（人） Number of Students (person)			
		毕业生 Graduates	招 生 Entrants	在校生 Enrollment	预计毕业生数 Estimated Graduates of Next Year
水声工程 Hydroacoustic Engineering	5	12	31	149	87
船舶与海洋工程新专业 New Speciality of Shipand Marine Engineering	3	4	4	20	13
捕捞学 Science of Fishing	2	0	3	15	11
水产品加工及贮藏工程 Aquatic Product Sprocessing and Storing Engineering	9	5	8	53	31
水产新专业 New Specialities of Aquaculture	3	10	7	51	33
水产养殖 Aquaculture	7	37	76	237	107
水生生物学 Hydrobiology	19	87	109	375	175
水文学及水资源 Hydrology and Water Resource	18	90	118	620	359
渔业资源 Fishery Resource	4	12	15	63	38

7-2 全国各海洋专业硕士研究生情况
Postgraduate Students from Marine Specialities

专 业 Speciality	专业点数（个） Number of Speciality Agencies (unit)	学生数（人） Number of Students (person)			
		毕业生 Graduates	招 生 Entrants	在校生 Enrollment	预计毕业生数 Estimated Graduates of Next Year
合 计 Total	**327**	**3 217**	**3 179**	**9 999**	**3 531**
物理海洋学 Physical Oceanography	12	80	123	339	101
海洋化学 Marine Chemistry	15	90	116	312	100
海洋生物学 Marine Biology	25	365	331	1 128	389
海洋地质 Marine Geology	20	111	155	452	147
海洋科学新专业 New Speciality of Marine Science	3	53	60	189	59
港口海岸、及近海工程 Coastal Harbour and Offshore Engineering	24	315	274	887	324
船舶与海洋结构物设计制造 Ship and Marine Structures Design and Manufacture	21	368	375	1 115	386
轮机工程 Turbine Engineering	14	267	233	728	317

7-2 续表 continued

专 业 Speciality	专业点数（个） Number of Speciality Agencies (unit)	学生数（人） Number of Students (person)			
		毕业生 Graduates	招 生 Entrants	在校生 Enrollment	预计毕业生数 Estimated Graduates of Next Year
水声工程 Hydroacoustic Engineering	11	128	123	363	138
船舶与海洋工程新专业 New Specialities in Shipping and Marine Engineering	8	15	40	73	15
捕捞学 Science of Fishing	4	32	26	88	30
航空、航天与航海医学 Aeronautical, Aerospace and Nautical Medicine	2	15	14	45	15
水产品加工及贮藏工程 Aquatic Products Processing and Storing Engineering	25	75	106	292	91
水产新专业 New Specialities of Aquaculture	3	18	14	44	13
水产养殖 Aquaculture	30	405	417	1 303	445
水生生物学 Hydrobiology	44	364	284	1 041	368
水文学及水资源 Hydrology and Water Resource	56	445	402	1 337	508
渔业资源 Fishery Resource	10	71	86	263	85

7-3 全国普通高等教育各海洋专业本科学生情况
Undergraduates from Marine Specialities in the National Ordinary Higher Education

专业 Speciality	专业点数（个） Number of Speciality Agencies (unit)	学生数（人） Number of Students (person)			
		毕业生 Graduates	招生 Entrants	在校生 Enrollment	预计毕业生数 Estimated Graduates of Next Year
合计 Total	**211**	**13 596**	**15 773**	**61 325**	**15 226**
海洋科学 Marine Science	21	701	699	3 156	853
海洋技术 Marine Technology	16	519	502	2 098	548
海洋管理 Marine Management	3	99	79	313	84
海洋生物资源与环境 Marine Living Resources and Environment	8	194	238	719	207
海洋科学新专业 New Specialties Under the Category of Marmine Sciences	3	0	137	221	0
港口海岸及治河工程 Coastal Harbour and River Control Engineering	1	60	0	122	58
港口航道与海岸工程 Harbour Channel and Coastal Engineering	24	1 410	1 577	6 694	1 753
水资源与海洋工程 Water Resource and Ocean Engineering	2	13	31	133	23
航海技术 Nautical Technology	16	2 278	2 570	9 783	2 500
轮机工程 Turbine Engineering	20	2 503	3 024	11 576	2 923
海事管理 Maritime Affairs Manegernent	2	120	122	485	115
船舶与海洋工程 Ship and Marine Engineering	29	2 956	2 706	11 927	3 067
海洋工程类新专业 New Speciality of Marine Engineering	1	0	197	396	0
水产养殖学 Aquaculture	48	2 213	3 010	10 716	2 482
海洋渔业科学与技术 Science and Technology of Marine Fishery	9	323	509	1 684	411
水族科学与技术 Science and Technology of Aquatic Animals	6	207	295	992	202
水产类新专业 New Speciality in Aquatic Products	2	0	77	310	0

7-4 全国普通高等教育各海洋专业专科学生情况
Students from Marine Specialities of the Colleges for Professional Training in the National Ordinary Higher Education

专 业 Speciality	专业点数（个） Number of Speciality Agencies (unit)	学生数（人） Number of Students (person)			
		毕业生 Graduates	招 生 Entrants	在校生 Enrollment	预计毕业生数 Estimated Graduates of Next Year
合 计 Total	**464**	**36 238**	**30 132**	**101 041**	**36 158**
港口工程技术 Harbour Engineering	9	518	611	1 729	576
港口业务管理 Harbour Business Management	18	883	874	3 170	1 104
港口与航运管理 Harbour and Shipping Management	10	314	468	1 257	349
港口机械应用技术 Applied Harbour Machinery Technology	2	162	285	742	237
港口物流设备与自动控制 Harbour Logistics Equipment and Automatic Control	20	1 270	1 054	3 177	1 075
集装箱运输管理 Container Transportation Management	19	996	652	2 539	971
国际航运业务管理 International Shipping Business Management	31	1 898	2 249	6 209	1 934
海关管理 Customs Management	5	205	61	88	14
报关与国际货运 Customs Clearing and International Freight Transport	194	15 964	11 932	41 011	15 387
轮机工程技术 Engine Engineering Technology	44	5 872	5 090	17 655	6 430
船舶检验 Ship Inspection	5	162	75	356	99
船舶舾装 Ship Equipment and Installations	5	154	105	408	115

7-4 续表 continued

专 业 Speciality	专业点数（个） Number of Speciality Agencies (unit)	学生数（人） Number of Students (person)			
		毕业生 Graduates	招 生 Entrants	在校生 Enrollment	预计毕业生数 Estimated Graduates of Next Year
船艇动力管理 Ship and Boat Power Equipment Management	1	80	0	48	0
船舶工程技术 Ship Engineering Technology	38	5 523	4 518	15 997	5 612
船机制造与维修 Ship Engines Manufacturing and Maintenance	7	201	406	1 201	317
水运管理 Water Transport Management	3	54	116	379	120
水信息技术 Hydrological Information Technology	2	19	17	94	58
水环境监测与保护 Water Environmental Monitoring and Protection	7	264	147	625	285
水政水资源管理 Water Resources Management by Water Administration	2	44	102	246	49
水产养殖技术 Aquaculture Technology	31	1 306	1 251	3 499	1 117
渔业综合技术 Integrated Fishery atechnology	3	108	1	78	65
水生动植物保护 Aquatic Animals and Plants Conservation	2	74	0	83	66
水上运输类新专业 New Specialities in Waterborne Communications	1	0	0	1	1
港口航道与治河工程 Harbour, Channel and River Harnessing Engineering	4	167	100	431	177
水文自动化测报技术 New Specialties Under the Category of Marmine Sciences	1	0	18	18	0

7-5 全国成人高等教育各海洋专业本科学生情况
Undergraduates from Marine Specialities in the National Adult Higher Education

专 业 Speciality	专业点数（个） Number of Speciality Agencies (unit)	学生数（人） Number of Students (person)			
		毕业生 Graduates	招 生 Entrants	在校生 Enrollment	预计毕业生数 Estimated Graduates of Next Year
合 计 Total	**54**	**2 214**	**2 504**	**6 122**	**2 176**
生物科学 Bioscience	4	64	46	100	54
生物技术（可授理学、工学学位） Biotechnology (The speciality may confer-degrees of science and engineering.)	2	75	137	443	210
港口航道与海岸工程 Harbour Channel and Coastal Engineering	2	44	25	109	25
航海技术 Navigation Technology	8	167	178	765	332
轮机工程 Turbine Engineering	8	253	388	793	264
海事管理 Maritime Affairs Management	2	34	28	52	24
船舶与海洋工程 Ship and Marine Engineering	14	1 402	1 506	3 354	1 121
水产养殖学 Aquaculture	14	175	196	506	146

7-6 全国成人高等教育各海洋专业专科学生情况
Students from Marine Specialities of the Colleges for Professional Training in the National Adult Higher Education

专 业 Speciality	专业点数（个） Number of Speciality Agencies (unit)	学生数（人） Number of Students (person)			
		毕业生 Graduates	招 生 Entrants	在校生 Enrollment	预计毕业生数 Estimated Graduates of Next Year
合 计 Total	**136**	**6 679**	**6 768**	**18 839**	**8 283**
港口业务管理 Harbour Business Management	3	63	0	12	12
港口与航运管理 Harbour and Shipping Management	1	14	1	2	1
国际航运业务管理 International Shipping Business Management	4	42	38	150	41
港口物流设备与自动控制 Harbour Logistics Equipment and Automatic Control	1	94	52	198	146
轮机工程技术 Engine Engineering	31	3 143	4 160	10 776	4 635
船舶检验 Ship Inspection	2	13	7	52	38
船舶工程技术 Ship Engineering	20	1 974	1 425	4 475	2 146
船机制造与维修 Ship Engines Manufacturing and Maintenance	2	8	39	112	1

7-6 续表 continued

专 业 Speciality	专业点数（个） Number of Speciality Agencies (unit)	学生数（人） Number of Students (person)			
		毕业生 Graduates	招 生 Entrants	在校生 Enrollment	预计毕业生数 Estimated Graduates of Next Year
船艇动力管理 Ship and Boat Power Equipment Management	1	0	0	79	31
水上运输类新专业 New Specialties Under the category of Water Transport	1	47	45	81	36
海关管理 Customs Management	1	0	0	1	0
报关与国际货运 Customs Declaration/International Freight Transport	38	1 074	842	2 424	1 028
水政水资源管理 Water Resources Management by Water Administration	4	21	21	70	34
水运管理 Water Transport Management	1	11	0	0	0
水文与水资源类新专业 New Specialities in Hydrology and Water Resources	1	0	0	2	2
水产养殖技术 Aquaculture Technology	23	172	135	401	131
水产养殖类新专业 New Specialties Under the category of Aquaculture	1	0	3	4	1
港口航道与治河工程 Harbour Channel and River Harnessing Projects	1	3	0	0	0

7-7 全国中等职业教育各海洋专业学生情况
Students from Marine Specialities in the National Secondary Vocational Education

专 业 Speciality	专业点数（个） Number of Speciality Agencies (unit)	学生数（人） Number of Students (person)			
		毕业生 Graduates	招 生 Entrants	在校生 Enrollment	预计毕业生数 Estimated Graduates of Next Year
合 计 Total	**459**	**43 561**	**24 127**	**71 337**	**29 130**
海水生态养殖 Seawater Ecological Cultivation	25	654	1 271	4 764	1 564
农林牧渔类新专业 New Specialities of Agriculture, Forestry, Animal Husbaudry and Fishery	60	15 080	3 134	9 669	5 035
航海捕捞 Sea Fishing	9	254	70	536	310
水文与水资源勘测 Hydrological and Water Resources Survey	3	25	139	327	132
风电场机电设备运行与维护 Operation and Maintenance of Electromechanical Equipment in	45	2 463	2 298	7 617	2 701
船舶制造与修理 Ships Building and Repair	60	2 701	1 818	7 521	2 555
船舶机械装置安装与维修 Installation and Maintenance of Ships' Mechanical Equipment	15	265	66	363	163
船舶驾驶 Ship Piloting	91	11 937	6 789	19 237	8 358

7-7 续表 continued

专 业 Speciality	专业点数（个） Number of Speciality Agencies (unit)	学生数（人） Number of Students (person)			
		毕业生 Graduates	招 生 Entrants	在校生 Enrollment	预计毕业生数 Estimated Graduates of Next Year
轮机管理 Engines Management	60	7 852	5 787	13 113	5 567
船舶水手与机工 Ship Sailors and Mechanics	35	766	1 058	2 754	921
船舶电气技术 Ship Electric Technology	21	483	245	1 615	593
外轮理货 Foreign Ships Freight Forwarding	13	589	197	960	433
船舶检验 Ships Inspection	2	80	0	39	26
港口机械运行与维护 Operation and Maintenance of Harbour Machinery	18	394	1 226	2 772	751
工程潜水 Engineering Diving	2	18	29	50	21

注：此表不包含技工学校相关数据。

Note: Data related to Technical Schools are not included in the table.

7-8 分地区各海洋专业博士研究生情况
Doctoral Students in Marine Specialities by Regions

地 区 Region	专业点数（个） Number of Speciality Agencies (unit)	学生数（人） Number of Students (person)			
		毕业生 Graduates	招 生 Entrants	在校生 Enrollment	预计毕业生数 Estimated Graduates of Next Year
合 计 Total	**131**	**615**	**870**	**3 660**	**1 806**
北 京 Beijing	7	24	25	113	54
天 津 Tianjin	4	6	1	5	4
辽 宁 Liaoning	5	41	46	312	46
上 海 Shanghai	14	35	45	211	104
江 苏 Jiangsu	14	49	77	428	274
浙 江 Zhejiang	4	4	23	59	22
福 建 Fujian	6	19	32	121	60
山 东 Shandong	19	231	275	1 087	581
广 东 Guangdong	13	62	76	262	118
海 南 Hainan	1	0	1	1	0
其 他 Others	44	144	269	1 061	543

7-9　分地区各海洋专业硕士研究生情况
Postgraduate Students in Marine Specialities by Regions

地　区 Region	专业点数（个） Number of Speciality Agencies (unit)	学生数（人） Number of Students (person)			
		毕业生 Graduates	招 生 Entrants	在校生 Enrollment	预计毕业生数 Estimated Graduates of Next Year
合　计 Total	**327**	**3 217**	**3 179**	**9 999**	**3 531**
北　京 Beijing	20	100	87	299	113
天　津 Tianjin	8	38	43	116	36
河　北 Hebei	5	20	4	38	17
辽　宁 Liaoning	21	379	378	1 095	415
上　海 Shanghai	27	358	275	944	366
江　苏 Jiangsu	24	367	368	1 168	400
浙　江 Zhejiang	21	170	125	463	156
福　建 Fujian	19	137	219	572	156
山　东 Shandong	36	419	474	1 463	491
广　东 Guangdong	30	242	295	835	275
广　西 Guangxi	4	24	17	62	23
海　南 Hainan	3	24	23	79	29
其　他 Others	109	939	871	2 865	1 054

7-10 分地区普通高等教育各海洋专业本科学生情况
Undergraduates in the Marine Specialities of Ordinary Higher Education by Regions

地 区 Region	专业点数（个） Number of Speciality Agencies (unit)	学生数（人） Number of Students (person)			
		毕业生 Graduates	招 生 Entrants	在校生 Enrollment	预计毕业生数 Estimated Graduates of Next Year
合 计 Total	**211**	**13 596**	**15 773**	**61 325**	**15 226**
北 京 Beijing	3	91	97	566	163
天 津 Tianjin	12	679	903	3 426	789
河 北 Hebei	7	211	187	805	235
辽 宁 Liaoning	20	2 100	1 965	8 333	2 161
上 海 Shanghai	13	1 240	1 187	5 373	1 569
江 苏 Jiangsu	28	1 454	1 438	5 779	1 469
浙 江 Zhejiang	21	712	705	3 012	670
福 建 Fujian	10	968	1 557	4 657	985
山 东 Shandong	25	1 542	2 156	7 560	1 972
广 东 Guangdong	14	775	1 064	3 751	894
广 西 Guangxi	6	136	238	754	147
海 南 Hainan	2	101	157	507	104
其 他 Others	50	3 587	4 119	16 802	4 068

7-11 分地区普通高等教育各海洋专业专科学生情况
Students from of the Marine Specialities Colleges for Professional Training in the Ordinary Higher Education by Regions

地 区 Region	专业点数（个） Number of Speciality Agencies (unit)	学生数（人） Number of Students (person)			
		毕业生 Graduates	招 生 Entrants	在校生 Enrollment	预计毕业生数 Estimated Graduates of Next Year
合 计 Total	**464**	**36 238**	**30 132**	**101 041**	**36 158**
天 津 Tianjin	13	1 614	1 354	4 533	1 688
河 北 Hebei	24	1 529	440	2 602	1 320
辽 宁 Liaoning	31	2 338	3 633	9 120	2 506
上 海 Shanghai	30	2 483	1 652	6 026	2 381
江 苏 Jiangsu	53	5 707	3 453	14 281	5 503
浙 江 Zhejiang	30	1 905	1 970	6 072	2 113
福 建 Fujian	28	2 086	1 654	5 435	1 980
山 东 Shandong	65	5 868	5 028	17 177	6 213
广 东 Guangdong	31	1 773	1 942	5 902	1 828
广 西 Guangxi	25	1 411	1 199	3 717	1 281
海 南 Hainan	9	602	474	1 339	515
其 他 Others	125	8 922	7 333	24 837	8 830

7-12 分地区成人高等教育各海洋专业本科学生情况
Students from Marine Specialities in the Adult Higher Education by Regions

地 区 Region	专业点数（个） Number of Speciality Agencies (unit)	学生数（人） Number of Students (person)			
		毕业生 Graduates	招 生 Entrants	在校生 Enrollment	预计毕业生数 Estimated Graduates of Next Year
合 计 Total	**54**	**2 214**	**2 504**	**6 122**	**2 176**
北 京 Beijing	1	75	137	340	107
天 津 Tianjin	2	66	50	200	78
河 北 Hebei	1	26	9	47	18
辽 宁 Liaoning	8	293	288	734	446
上 海 Shanghai	2	72	63	234	140
江 苏 Jiangsu	3	1 087	1 051	2 355	830
浙 江 Zhejiang	6	62	131	366	62
福 建 Fujian	1	3	16	53	16
山 东 Shandong	9	136	62	312	136
广 东 Guangdong	4	0	205	507	27
广 西 Guangxi	1	18	13	32	19
其 他 Others	16	376	479	942	297

7-13 分地区成人高等教育各海洋专业专科学生情况
Students from Marine Specialities of the colleges for Professional Training in the Adult Higher Education by Regions

地 区 Region	专业点数（个） Number of Speciality Agencies (unit)	学生数（人） Number of Students (person)			
		毕业生 Graduates	招 生 Entrants	在校生 Enrollment	预计毕业生数 Estimated Graduates of Next Year
合 计 Total	**136**	**6 679**	**6 768**	**18 839**	**8 283**
天 津 Tianjin	3	111	168	369	99
河 北 Hebei	2	85	0	34	23
辽 宁 Liaoning	12	1 624	1 462	4 334	2 581
上 海 Shanghai	16	478	565	1 843	592
江 苏 Jiangsu	19	1 317	1 583	2 946	1 149
浙 江 Zhejiang	12	1 129	965	2 931	1 413
福 建 Fujian	5	210	156	649	266
山 东 Shandong	14	163	402	1 385	498
广 东 Guangdong	11	526	430	1 146	661
广 西 Guangxi	4	2	7	9	2
海 南 Hainan	1	1	1	8	6
其 他 Others	37	1 033	1 029	3 185	993

7-14 分地区中等职业教育各海洋专业学生情况
Students from Marine Specialities in the Secondary Vocational Education by Regions

地 区 Region	专业点数（个） Number of Speciality Agencies (unit)	学生数（人） Number of Students (person)			
		毕业生 Graduates	招 生 Entrants	在校生 Enrollment	预计毕业生数 Estimated Graduates of Next Year
合 计 Total	**459**	**43 561**	**24 127**	**71 337**	**29 130**
天 津 Tianjin	6	687	341	926	553
河 北 Hebei	35	6 461	2 531	6 372	2 725
辽 宁 Liaoning	50	1 564	2 654	5 538	1 466
上 海 Shanghai	11	351	244	903	247
江 苏 Jiangsu	52	1 939	875	6 833	2 802
浙 江 Zhejiang	26	1 742	1 323	4 150	1 582
福 建 Fujian	43	4 501	3 628	8 047	3 238
山 东 Shandong	57	10 129	5 001	14 037	5 936
广 东 Guangdong	21	1 478	613	2 742	1 224
广 西 Guangxi	11	306	1 239	3 283	409
海 南 Hainan	3	199	236	525	115
其 他 Others	144	14 204	5 442	17 981	8 833

7-15 分地区开设海洋专业高等学校教职工数
Number of Teaching and Administrative Staff in the Universities and Colleges Offering Marine Specialities by Regions

地 区 Region	学校（机构）数（个） Number of Colleges (Institutions) (unit)	教职工数（人） Number of Teaching and Administrative Staff (person)	专任教师数（人） Number of Full-Time Teachers (person)
合 计 Total	**361**	**435 906**	**271 996**
北 京 Beijing	3	5 390	3 230
天 津 Tianjin	12	15 629	10 332
河 北 Hebei	19	17 265	10 270
辽 宁 Liaoning	21	18 230	11 253
上 海 Shanghai	16	21 234	10 726
江 苏 Jiangsu	41	66 227	42 194
浙 江 Zhejiang	20	25 246	14 427
福 建 Fujian	16	13 388	9 234
山 东 Shandong	46	52 961	34 188
广 东 Guangdong	21	30 135	18 741
广 西 Guangxi	16	13 381	9 691
海 南 Hainan	6	5 894	3 677
其 他 Others	124	150 926	94 033

主要统计指标解释

海洋专业　指高等教育和中等职业教育所设的与海洋有关的专业。

Explanatory Notes on Main Statistical Indicators

Marine Speciality refers to the ocean-related speciality in high learning and the professional secondary vocational education.

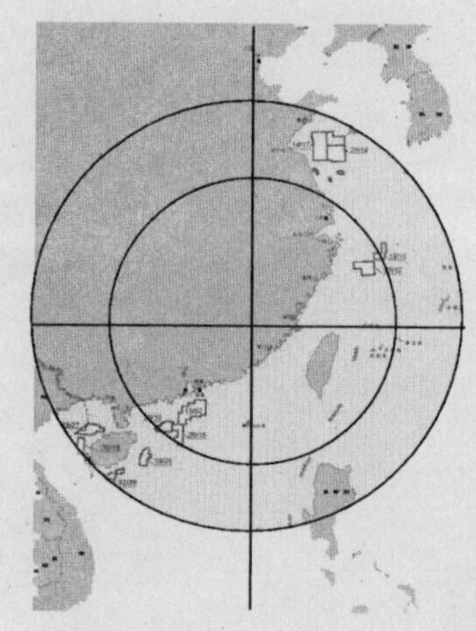

8

海洋环境保护

Marine Environmental Protection

8-1 海区海水水质评价结果
Seawater Quality Assessment Results

项 目 Item	**全国总计 National Total**	渤 海 Bohai Sea	黄 海 Huanghai Sea	东 海 East China Sea	南 海 South China Sea
合计 （平方千米） Total (km^2)	**169 520**	41 140	40 410	63 130	24 840
第二类水质海域面积 （平方千米） Area of the Second Grade Sea Waters (km^2)	**46 910**	12 330	12 890	12 800	8 890
第三类水质海域面积 （平方千米） Area of the Third Grade Sea Waters (km^2)	**30 030**	11 040	3 450	7 540	8 000
第四类水质海域面积 （平方千米） Area of the Fourth Grade Sea Waters (km^2)	**24 700**	4 690	7 540	8 820	3 650
劣于第四类海域面积 （平方千米） Sea Area of the Sea Waters Inferior to the Fourth Grade (km^2)	**67 880**	13 080	16 530	33 970	4 300
首要超标污染物 Prime Pollutants Exceeding the Set Standard	**无机氮、活性磷酸盐、石油类 Inorganic Nitrogen, Active Phosphate, Oils**	无机氮、石油类 Inorganic Nitrogen, Oils	无机氮、活性磷酸盐、石油类 Inorganic Nitrogen, Active Phosphate, Oils	无机氮、活性磷酸盐 Inorganic Nitrogen, Active Phosphate	无机氮、活性磷酸盐、石油类 Inorganic Nitrogen, Active Phosphate, Oils

8-2 海区废弃物海洋倾倒情况
Ocean Dumping of Wastes by Sea Area

海 区 Sea Area	疏浚物 （万立方米） Dredged Materials (10 000m³)	惰性无机地质废料 （立方米） Inert, Inorganic Geologic Wastes (m³)
合 计 **Total**	**20 029.01**	**9 447 400**
渤黄海 Bohai and Huanghai Sea	4 738.00	
东 海 East China Sea	10 755.27	
南 海 South China Sea	4 535.74	9 447 400

8-3 海区海洋石油勘探开发污染物排放入海情况
Discharge of Pollutants into the Sea from Offshore Oil Exploration and Exploitation

海 区 Sea Area	生产污水（万立方米） Production Sewage (10 000m^3)	泥浆（立方米） Sludge (m^3)	钻屑（立方米） Debris from Drilling (m^3)	机舱污水（立方米） Sewage from Engineroom (m^3)	食品废弃物（立方米） Food Wastes (m^3)	生活污水（立方米） Domestic Sewage (m^3)
合 计 Total	**13 344.34**	**53 074.25**	**43 473.05**	**4 501.80**		**590 888.86**
渤 海 Bohai Sea	682.12	10 836.71	31 712.47		5.63*	284 232.76
黄 海 Huanghai Sea						
东 海 East China Sea	74.59	1 504.14	3 240.18		584.91*	34 508.00
南 海 South China Sea	12 587.63	40 733.40	8 520.40	4 501.80	579.71	272 148.10

注：*单位为吨。

Notes:* unit is ton.

8-4 沿海地区工业废水排放及处理情况
Discharge and Treatment of Industrial Waste Water by Coastal Regions

单位：万吨 (10 000 t)

地 区 Region		工业废水排放总量 Total Volume of Industrial Waste Water Discharged	直排入海 Discharged Directly into the Sea
总 计 Total		**1 281 013.37**	**107 106.45**
环渤海经济区 Round-the-Bohai Sea Economic Zone	**合 计 Total**	**412 563.06**	**36 474.60**
	辽 宁 Liaoning	87 167.54	26 579.43
	河 北 Hebei	122 644.54	1 478.56
	天 津 Tianjin	19 116.85	43.40
	山 东 Shandong	183 634.13	8 373.21
长江三角洲经济区 Yangtze River Delta Economic Zone	**合 计 Total**	**457 868.88**	**19 685.34**
	江 苏 Jiangsu	236 094.40	983.66
	上 海 Shanghai	46 358.62	10 190.62
	浙 江 Zhejiang	175 415.86	8 511.07
海峡西岸经济区 Economic Zone on the West Side of the Taiwan Straits	**合 计 Total**	**106 319.29**	**39 701.86**
	福 建 Fujian	106 319.29	39 701.86
珠江三角洲经济区 Zhujiang River Delta Economic Zone	**合 计 Total**	**186 126.32**	**6 332.77**
	广 东 Guangdong	186 126.32	6 332.77
环北部湾经济区 Round-the-Beibu Gulf Economic Zone	**合 计 Total**	**118 135.83**	**4 911.88**
	广 西 Guangxi	110 670.98	1 584.10
	海 南 Hainan	7 464.85	3 327.78

8-5 沿海城市工业废水排放及处理情况
Discharge and Treatment of Industrial Waste Water by Coastal Cities

单位：万吨 (10 000 t)

沿海城市 Coastal City	工业废水排放总量 Total Volume of Industrial Waste Water Discharged	
		直排入海 Discharged Directly into the Sea
天　津 Tianjin	19 116.85	43.40
唐　山 Tangshan	19 396.37	1 059.03
秦皇岛 Qinhuangdao	6 055.23	0.00
沧　州 Cangzhou	11 665.81	137.82
大　连 Dalian	30 794.90	24 960.15
丹　东 Dandong	5 509.88	13.61
锦　州 Jinzhou	6 533.80	65.04
营　口 Yingkou	2 937.19	894.23
盘　锦 Panjin	3 935.96	0.00
葫芦岛 Huludao	2 656.26	610.56
上　海 Shanghai	46 358.62	10 190.62
南　通 Nantong	18 199.93	9.15
连云港 Lianyungang	6 390.42	495.59
盐　城 Yancheng	20 117.27	442.00

8-5 续表1 continued

沿海城市 Coastal City	工业废水排放总量 Total Volume of Industrial Waste Water Discharged	
		直排入海 Discharged Directly into the Sea
杭　州 Hangzhou	42 723.67	0.00
宁　波 Ningbo	20 124.65	6 978.19
温　州 Wenzhou	7 912.99	11.29
嘉　兴 Jiaxing	23 266.70	232.20
绍　兴 Shaoxing	30 417.55	0.00
舟　山 Zhoushan	1 950.91	1 122.66
台　州 Taizhou	6 151.64	166.73
福　州 Fuzhou	5 332.67	144.25
厦　门 Xiamen	26 947.88	22 454.57
莆　田 Putian	2 262.22	137.54
泉　州 Quanzhou	20 534.98	2 106.06
漳　州 Zhangzhou	23 400.07	14 753.39
宁　德 Ningde	1 586.98	84.83
青　岛 Qingdao	11 145.96	1 580.18
东　营 Dongying	10 153.07	0.00
烟　台 Yantai	9 358.92	2 607.86
潍　坊 Weifang	26 954.80	0.57
威　海 Weihai	2 478.87	94.95
日　照 Rizhao	7 868.18	3 523.07
滨　州 Binzhou	16 830.69	0.00

8-5 续表2 continued

沿海城市 Coastal City	工业废水排放总量 Total Volume of Industrial Waste Water Discharged	
		直排入海 Discharged Directly into the Sea
广　州　Guangzhou	22 676.54	19.00
深　圳　Shenzhen	13 830.52	623.02
珠　海　Zhuhai	5 523.96	57.52
汕　头　Shantou	5 097.44	41.11
江　门　Jiangmen	14 413.08	68.17
湛　江　Zhanjiang	8 263.40	739.01
茂　名　Maoming	5 205.56	997.05
惠　州　Huizhou	8 300.29	980.00
汕　尾　Shanwei	1 902.24	4.87
阳　江　Yangjiang	2 045.76	1.60
东　莞　Dongguan	26 909.03	2 314.38
中　山　Zhongshan	8 622.35	0.00
潮　州　Chaozhou	2 993.97	0.00
揭　阳　Jieyang	3 742.72	0.00
北　海　Beihai	2 290.74	239.33
防城港　Fangchenggang	3 108.31	870.99
钦　州　Qinzhou	1 920.90	473.78
海　口　Haikou	879.10	0.00
三　亚　Sanya	91.93	0.00

8-6 沿海地带工业废水排放及处理情况
Discharge and Treatment of Industrial Waste Water by Coastal Counties

单位：万吨 (10 000 t)

地区 Region	工业废水排放总量 Total Volume of Industrial Waste Water Discharged	直排入海 Discharged Directly into the Sea
天津 Tianjin	**9 532.21**	**43.40**
河北 Hebei	**7 059.06**	**0.00**
唐山 Tangshan	2 235.64	0.00
秦皇岛 Qinhuangdao	3 556.08	0.00
沧州 Cangzhou	1 267.34	0.00
辽宁 Liaoning	**36 504.97**	**23 033.94**
大连 Dalian	26 343.43	21 515.54
丹东 Dandong	1 741.04	13.61
锦州 Jinzhou	2 996.69	0.00
营口 Yingkou	2 146.62	894.23
盘锦 Panjin	808.06	0.00
葫芦岛 Huludao	2 469.14	610.56
上海 Shanghai	**31 783.66**	**10 190.62**
江苏 Jiangsu	**24 921.79**	**937.59**
南通 Nantong	9 291.45	0.00
连云港 Lianyungang	4 318.09	495.59
盐城 Yancheng	11 312.25	442.00

8-6 续表1 continued

地 区 Region	工业废水排放总量 Total Volume of Industrial Waste Water Discharged	直排入海 Discharged Directly into the Sea
浙 江 Zhejiang	**74 807.21**	**8 306.50**
杭 州 Hangzhou	13 276.38	0.00
宁 波 Ningbo	17 827.82	6 947.72
温 州 Wenzhou	5 832.07	11.29
嘉 兴 Jiaxing	9 289.73	58.10
绍 兴 Shaoxing	22 394.68	0.00
舟 山 Zhoushan	1 950.91	1 122.66
台 州 Taizhou	4 235.63	166.73
福 建 Fujian	**70 591.43**	**39 680.64**
福 州 Fuzhou	4 153.92	144.25
厦 门 Xiamen	26 947.88	22 454.57
莆 田 Putian	2 262.22	137.54
泉 州 Quanzhou	17 835.91	2 106.06
漳 州 Zhangzhou	18 331.58	14 753.39
宁 德 Ningde	1 059.90	84.83
山 东 Shandong	**45 387.12**	**4 654.33**
青 岛 Qingdao	9 698.78	1 580.18
东 营 Dongying	9 767.66	0.00
烟 台 Yantai	7 822.62	2 607.86
潍 坊 Weifang	10 778.97	93.23
威 海 Weihai	1 916.92	0.00
日 照 Rizhao	1 182.91	373.07
滨 州 Binzhou	4 219.27	0.00

8-6 续表2 continued

地 区 Region	工业废水排放总量 Total Volume of Industrial Waste Water Discharged	直排入海 Discharged Directly into the Sea
广 东 Guangdong	**57 456.45**	**1 504.29**
广 州 Guangzhou	11 614.04	19.00
深 圳 Shenzhen	12 041.05	623.02
珠 海 Zhuhai	2 800.73	34.88
汕 头 Shantou	5 055.55	41.11
江 门 Jiangmen	9 226.66	68.17
湛 江 Zhanjiang	7 988.72	695.64
茂 名 Maoming	2 107.62	16.06
惠 州 Huizhou	1 531.14	0.00
汕 尾 Shanwei	1 781.08	4.80
阳 江 Yangjiang	1 296.33	1.60
东 莞 Dongguan	0.00	0.00
中 山 Zhongshan	0.00	0.00
潮 州 Chaozhou	1 202.14	0.00
揭 阳 Jieyang	811.40	0.00
广 西 Guangxi	**5 469.63**	**1 584.10**
北 海 Beihai	2 290.74	239.33
防城港 Fangchenggang	2 053.38	870.99
钦 州 Qinzhou	1 125.51	473.78
海 南 Hainan	**808.79**	**0.00**
海 口 Haikou	808.79	0.00
三 亚 Sanya	0.00	0.00

注：1. 表中数据为沿海地带合计数（表8-9、8-13同）。
2. 沿海地带是指有海岸线的县、县级市、区（包括直辖市和地级市的区）。

Note: 1. The data in the table are the total numbers for the coastal regions (The same for Tables 8-9, 8-13).
2.Coastal Zone refers to the counties, county-level cities and districts with coastlines (including the districts under the municipalities directly under the Central Government and the prefecture-level districts)

8-7 沿海地区一般工业固体废物倾倒丢弃、处理及综合利用情况
Discharge, Treatment and Multipurpose Utilization of Common Industrial Solid Wastes by Coastal Regions

单位：万吨　　　　(10 000 t)

地区 Region		一般工业固体废物倾倒丢弃量 Volume of Common Industrial Solid Wastes Discharged	一般工业固体废物处置量 Volume of Common Industrial Solid Wastes Treated	一般工业固体废物综合利用量 Volume of Common Industrial Solid Wastes Comprehansively Utilized
总 计 Total		**14.802**	**25 020.38**	**81 369.80**
环渤海经济区 Round-the-Bohai Sea Economic Zone	**合 计 Total**	**10.406**	**20 162.19**	**48 112.01**
	辽 宁 Liaoning	10.402	11 654.99	11 861.83
	河 北 Hebei	0.000	7 439.07	17 360.83
	天 津 Tianjin	0.000	6.80	1 816.50
	山 东 Shandong	0.004	1 061.33	17 072.86
长江三角洲经济区 Yangtze River Delta Economic Zone	**合 计 Total**	**0.655**	**1 000.61**	**15 564.69**
	江 苏 Jiangsu	0.013	630.40	9 341.57
	上 海 Shanghai	0.247	55.86	2 140.36
	浙 江 Zhejiang	0.395	314.35	4 082.76
海峡西岸经济区 Economic Zone on the West Side of the Taiwan Straits	**合 计 Total**	**0.163**	**764.27**	**6 887.40**
	福 建 Fujian	0.163	764.27	6 887.40
珠江三角洲经济区 Zhujiang River Delta Economic Zone	**合 计 Total**	**3.117**	**814.48**	**5 198.31**
	广 东 Guangdong	3.117	814.48	5 198.31
环北部湾经济区 Round-the-Beibu Gulf Economic Zone	**合 计 Total**	**0.461**	**2 278.83**	**5 607.39**
	广 西 Guangxi	0.407	2 218.85	5 369.24
	海 南 Hainan	0.054	59.98	238.15

8-8 沿海城市一般工业固体废物倾倒丢弃、处理及综合利用情况
Discharge, Treatment and Multipurposed Utilization of Common Industrial Solid Wastes by Coastal Cities

单位：万吨 (10 000 t)

沿海城市 Coastal City	一般工业固体废物 倾倒丢弃量 Volume of Common Industrial Solid Wastes Discharged	一般工业固体废物 处置量 Volume of Common Industrial Solid Wastes Treated	一般工业固体废物 综合利用量 Volume of Common Industrial Solid Wastes Comprehansively Utilized
天　津　Tianjin	0.000	6.80	1 816.50
唐　山　Tangshan	0.000	2 534.94	7 494.01
秦皇岛　Qinhuangdao	0.000	678.68	1 253.11
沧　州　Cangzhou	0.000	0.77	389.97
大　连　Dalian	0.000	31.51	675.28
丹　东　Dandong	0.000	436.51	163.11
锦　州　Jinzhou	0.000	75.41	205.72
营　口　Yingkou	0.000	0.97	525.16
盘　锦　Panjin	0.000	12.11	146.09
葫芦岛　Huludao	1.043	6.58	355.58
上　海　Shanghai	0.247	55.86	2 140.36
南　通　Nantong	0.000	8.12	438.87
连云港　Lianyungang	0.013	2.80	443.88
盐　城　Yancheng	0.000	37.63	187.81

8-8 续表1 continued

沿海城市 Coastal City	一般工业固体废物 倾倒丢弃量 Volume of Common Industrial Solid Wastes Discharged	一般工业固体废物 处置量 Volume of Common Industrial Solid Wastes Treated	一般工业固体废物 综合利用量 Volume of Common Industrial Solid Wastes Comprehansively Utilized
杭　州 Hangzhou	0.019	50.97	655.81
宁　波 Ningbo	0.021	93.24	1 147.00
温　州 Wenzhou	0.000	9.37	233.59
嘉　兴 Jiaxing	0.000	24.37	495.00
绍　兴 Shaoxing	0.005	19.84	349.45
舟　山 Zhoushan	0.000	0.95	74.25
台　州 Taizhou	0.296	10.73	228.80
福　州 Fuzhou	0.000	65.56	655.15
厦　门 Xiamen	0.000	10.72	109.87
莆　田 Putian	0.000	0.17	50.95
泉　州 Quanzhou	0.000	19.78	685.28
漳　州 Zhangzhou	0.000	8.25	174.57
宁　德 Ningde	0.000	11.79	207.85
青　岛 Qingdao	0.000	20.07	841.02
东　营 Dongying	0.000	17.60	309.24
烟　台 Yantai	0.000	446.04	2 075.03
潍　坊 Weifang	0.000	23.37	749.21
威　海 Weihai	0.000	32.07	331.91
日　照 Rizhao	0.000	4.64	988.43
滨　州 Binzhou	0.004	7.89	593.67

8-8 续表2 continued

沿海城市 Coastal City	一般工业固体废物倾倒丢弃量 Volume of Common Industrial Solid Wastes Discharged	一般工业固体废物处置量 Volume of Common Industrial Solid Wastes Treated	一般工业固体废物综合利用量 Volume of Common Industrial Solid Wastes Comprehansively Utilized
广　州 Guangzhou	0.000	22.26	589.09
深　圳 Shenzhen	0.072	9.92	94.41
珠　海 Zhuhai	0.000	9.95	263.86
汕　头 Shantou	0.000	1.28	106.47
江　门 Jiangmen	0.000	29.90	229.49
湛　江 Zhanjiang	0.088	11.60	243.45
茂　名 Maoming	0.000	5.57	95.72
惠　州 Huizhou	0.000	0.78	177.60
汕　尾 Shanwei	0.000	0.28	76.94
阳　江 Yangjiang	0.255	0.09	225.91
东　莞 Dongguan	0.651	122.17	415.86
中　山 Zhongshan	0.995	10.56	91.72
潮　州 Chaozhou	0.000	0.14	112.30
揭　阳 Jieyang	0.000	0.01	82.00
北　海 Beihai	0.028	47.87	219.86
防城港 Fangchenggang	0.030	0.09	140.77
钦　州 Qinzhou	0.049	2.07	145.82
海　口 Haikou	0.000	0.51	5.64
三　亚 Sanya	0.000	0.00	2.40

8-9 沿海地带一般工业固体废物倾倒丢弃、处理及综合利用情况
Discharge, Treatment and Multipurposed Utilization of Common Industrial Solid Wastes by Coastal Counties

单位：万吨　　(10 000 t)

地　区 Region	一般工业固体废物倾倒丢弃量 Volume of Common Industrial Solid Wastes Discharged	一般工业固体废物处置量 Volume of Common Industrial Solid Wastes Treated	一般工业固体废物综合利用量 Volume of Common Industrial Solid Wastes Comprehansively Utilized
天　津　Tianjin	**0.000**	**2.88**	**2 258.28**
河　北　Hebei	**0.000**	**180.69**	**4 107.99**
唐　山　Tangshan	0.000	92.83	2 000.80
秦皇岛　Qinhuangdao	0.000	87.86	890.79
沧　州　Cangzhou	0.000	0.00	1 216.40
辽　宁　Liaoning	**0.248**	**94.48**	**53 039.20**
大　连　Dalian	0.000	24.28	24 717.65
丹　东　Dandong	0.000	27.46	1 741.04
锦　州　Jinzhou	0.000	0.00	2 996.69
营　口　Yingkou	0.000	0.76	1 767.85
盘　锦　Panjin	0.000	3.32	808.06
葫芦岛　Huludao	0.001	6.51	2 469.14
上　海　Shanghai	**0.247**	**32.15**	**18 538.78**
江　苏　Jiangsu	**0.013**	**39.96**	**19 819.98**
南　通　Nantong	0.000	3.24	6 785.79
连云港　Lianyungang	0.013	1.62	3 346.64
盐　城　Yancheng	0.000	35.10	9 687.55

8-9 续表1 continued

地 区 Region	一般工业固体废物倾倒丢弃量 Volume of Common Industrial Solid Wastes Discharged	一般工业固体废物处置量 Volume of Common Industrial Solid Wastes Treated	一般工业固体废物综合利用量 Volume of Common Industrial Solid Wastes Comprehensively Utilized
浙 江 Zhejiang	**0.061**	**132.47**	**29 487.80**
杭 州 Hangzhou	0.019	13.98	3 084.30
宁 波 Ningbo	0.021	90.09	12 754.68
温 州 Wenzhou	0.000	8.36	3 992.10
嘉 兴 Jiaxing	0.000	0.84	614.60
绍 兴 Shaoxing	0.000	7.56	4 696.73
舟 山 Zhoushan	0.000	0.95	1 389.62
台 州 Taizhou	0.021	10.69	2 955.77
福 建 Fujian	**0.000**	**87.16**	**59 260.26**
福 州 Fuzhou	0.000	54.98	3 178.69
厦 门 Xiamen	0.000	10.72	23 808.78
莆 田 Putian	0.000	0.17	2 262.22
泉 州 Quanzhou	0.000	13.94	11 190.24
漳 州 Zhangzhou	0.000	7.26	18 065.51
宁 德 Ningde	0.000	0.09	754.82
山 东 Shandong	**0.000**	**499.05**	**23 730.24**
青 岛 Qingdao	0.000	19.43	2 732.36
东 营 Dongying	0.000	3.55	7 961.03
烟 台 Yantai	0.000	440.89	5 678.10
潍 坊 Weifang	0.000	2.33	2 462.98
威 海 Weihai	0.000	31.53	525.36
日 照 Rizhao	0.000	0.07	781.37
滨 州 Binzhou	0.000	1.26	3 589.04

8-9 续表2 continued

地 区 Region	一般工业固体废物倾倒丢弃量 Volume of Common Industrial Solid Wastes Discharged	一般工业固体废物处置量 Volume of Common Industrial Solid Wastes Treated	一般工业固体废物综合利用量 Volume of Common Industrial Solid Wastes Comprehensively Utilized
广 东 Guangdong	**0.350**	**71.64**	**44 557.04**
广 州 Guangzhou	0.000	18.34	7 176.58
深 圳 Shenzhen	0.071	8.76	6 922.49
珠 海 Zhuhai	0.000	2.37	1 882.46
汕 头 Shantou	0.000	1.28	4 285.06
江 门 Jiangmen	0.000	23.91	8 122.08
湛 江 Zhanjiang	0.036	10.98	7 930.45
茂 名 Maoming	0.000	5.56	2 103.76
惠 州 Huizhou	0.000	0.04	1 496.56
汕 尾 Shanwei	0.000	0.28	1 781.08
阳 江 Yangjiang	0.243	0.03	922.54
东 莞 Dongguan	0.000	0.00	0.00
中 山 Zhongshan	0.000	0.00	0.00
潮 州 Chaozhou	0.000	0.10	1 122.59
揭 阳 Jieyang	0.000	0.01	811.40
广 西 Guangxi	**0.058**	**48.17**	**5 067.76**
北 海 Beihai	0.028	47.87	1 888.87
防城港 Fangchenggang	0.030	0.09	2 053.38
钦 州 Qinzhou	0.000	0.21	1 125.51
海 南 Hainan	**0.000**	**0.51**	**62.58**
海 口 Haikou	0.000	0.51	62.58
三 亚 Sanya	0.000	0.00	0.00

8-10 主要沿海城市工业废气排放及处理情况
Emission and Treatment of Industrial Waste Gas in Major Coastal Cities

城 市 City	工业废气排放量（万立方米） Industrial Waste Gas Emission (10 000 cu.m)	工业二氧化硫排放量（万吨） Industrial Sulphur Dioxide Emission (10 000 t)	工业氮氧化物排放量（万吨） Industrial Nitrogen Oxide Emission (10 000 t)	工业烟（粉）尘排放量（万吨） Industrial Soot (Dust) Emission (10 000 t)	生活二氧化硫排放量（吨） Household Sulphur Dioxide Emission (t)	生活氮氧化物排放量（吨） Household Nitrogen Oxide Emission (t)	生活烟尘排放量（吨） Household Soot Emission (t)
天 津 Tianjin	9 032	21.5	27.6	5.9	8 959	4 447	18 400
秦皇岛 Qinhuangdao	2 794	7.2	5.7	7.9	4 338	965	55 498
大 连 Dalian	2 576	11.5	11.6	5.2	16 366	2 403	9 612
上 海 Shanghai	13 361	19.3	28.5	6.4	34 754	19 138	15 454
连云港 Lianyungang	970	4.1	3.4	1.7	6 588	844	3 178
宁波 Ningbo	6 218	14.4	23.6	2.6	2 477	692	487
温州 Wenzhou	1 701	3.7	4.3	1.9	599	370	467
福 州 Fuzhou	3 206	7.6	8.7	3.7	1 279	169	547
厦 门 Xiamen	1 178	1.9	1.3	0.3	486	181	190
青 岛 Qingdao	2 266	7.3	7.5	2.7	27 059	2 601	9 795
烟 台 Yantai	2 745	8.7	8.0	3.2	13 574	3 479	10 168
深 圳 Shenzhen	2 137	1.0	3.0	0.5	488	885	300
珠 海 Zhuhai	1 373	3.0	5.0	1.1	21	44	23
汕 头 Shantou	673	2.8	2.6	0.5	455	117	55
湛 江 Zhanjiang	1 009	2.5	2.2	1.0	2 458	395	258
北 海 Beihai	722	1.3	1.6	0.7	1 240	141	395
海 口 Haikou	27	0.2	0.0	0.1	178	32	174

8-11 沿海地区污染治理项目情况
Pollution Treatment Projects in Coastal Regions

单位：个 (unit)

地 区 Region		当年安排施工项目 Arranged for Construction in the Year		当年竣工项目 Completed in the Year	
		治理废水 Treatment of Waste Water	治理固体废物 Treatment of Solid Wastes	治理废水 Treatment of Waste Water	治理固体废物 Treatment of Solid Wastes
总 计 Total		**1 036**	**115**	**1 124**	**110**
环渤海经济区 Round-the-Bohai Sea Economic Zone	**合 计 Total**	**278**	**25**	**318**	**25**
	辽 宁 Liaoning	16	6	33	6
	河 北 Hebei	35	1	45	2
	天 津 Tianjin	50	2	20	2
	山 东 Shandong	177	16	220	15
长江三角洲经济区 Yangtze River Delta Economic Zone	**合 计 Total**	**365**	**25**	**406**	**20**
	江 苏 Jiangsu	130	8	158	7
	上 海 Shanghai	39	6	41	6
	浙 江 Zhejiang	196	11	207	7
海峡西岸经济区 Economic Zone on the West Side of the Taiwan Straits	**合 计 Total**	**109**	**43**	**102**	**42**
	福 建 Fujian	109	43	102	42
珠江三角洲经济区 Zhujiang River Delta Economic Zone	**合 计 Total**	**169**	**6**	**185**	**9**
	广 东 Guangdong	169	6	185	9
环北部湾经济区 Round-the-Beibu Gulf Economic Zone	**合 计 Total**	**115**	**16**	**113**	**14**
	广 西 Guangxi	95	16	92	13
	海 南 Hainan	20	0	21	1

8-12 沿海城市污染治理项目情况
Pollution Treatment Projects in Coastal Cities

单位：个　　(unit)

沿海城市 Coastal City	当年安排施工项目 Arranged for Construction in the Year		当年竣工项目 Completed in the Year	
	治理废水 Treatment of Waste Water	治理固体废物 Treatment of Solid Wastes	治理废水 Treament of Waste Water	治理固体废物 Treatment of Solid Wastes
天　津　Tianjin	50	2	20	2
唐　山　Tangshan	6	0	10	1
秦皇岛　Qinhuangdao	1	0	1	0
沧　州　Cangzhou	0	0	1	0
大　连　Dalian	3	0	5	0
丹　东　Dandong	1	3	2	3
锦　州　Jinzhou	1	0	2	0
营　口　Yingkou	0	0	1	0
盘　锦　Panjin	0	1	0	1
葫芦岛　Huludao	2	1	1	1
上　海　Shanghai	39	6	41	6
南　通　Nantong	24	2	29	3
连云港　Lianyungang	4	1	10	1
盐　城　Yancheng	1	0	2	0

8-12 续表1 continued

沿海城市 Coastal City	当年安排施工项目 Arranged for Construction in the Year		当年竣工项目 Completed in the Year	
	治理废水 Treatment of Waste Water	治理固体废物 Treatment of Solid Wastes	治理废水 Treament of Waste Water	治理固体废物 Treatment of Solid Wastes
杭　州 Hangzhou	8	3	6	2
宁　波 Ningbo	16	2	28	1
温　州 Wenzhou	10	0	9	0
嘉　兴 Jiaxing	24	1	23	1
绍　兴 Shaoxing	38	3	39	2
舟　山 Zhoushan	5	0	5	0
台　州 Taizhou	7	0	7	0
福　州 Fuzhou	6	0	4	0
厦　门 Xiamen	11	0	15	0
莆　田 Putian	1	0	1	0
泉　州 Quanzhou	36	37	24	36
漳　州 Zhangzhou	6	0	9	0
宁　德 Ningde	1	0	2	0
青　岛 Qingdao	4	1	4	1
东　营 Dongying	12	4	14	4
烟　台 Yantai	17	5	22	5
潍　坊 Weifang	38	0	36	0
威　海 Weihai	0	0	0	0
日　照 Rizhao	3	0	5	0
滨　州 Binzhou	5	0	8	0

8-12 续表2 continued

沿海城市 Coastal City	当年安排施工项目 Arranged for Construction in the Year		当年竣工项目 Completed in the Year	
	治理废水 Treatment of Waste Water	治理固体废物 Treatment of Solid Wastes	治理废水 Treament of Waste Water	治理固体废物 Treatment of Solid Wastes
广　州 Guangzhou	9	0	16	0
深　圳 Shenzhen	4	0	4	0
珠　海 Zhuhai	5	1	4	1
汕　头 Shantou	2	0	7	0
江　门 Jiangmen	21	0	24	0
湛　江 Zhanjiang	8	0	8	0
茂　名 Maoming	2	0	2	0
惠　州 Huizhou	4	0	5	0
汕　尾 Shanwei	1	1	0	0
阳　江 Yangjiang	0	0	0	0
东　莞 Dongguan	27	1	26	0
中　山 Zhongshan	2	0	2	0
潮　州 Chaozhou	1	0	2	0
揭　阳 Jieyang	9	0	10	0
北　海 Beihai	3	1	7	1
防城港 Fangchenggang	2	0	1	0
钦　州 Qinzhou	13	5	7	2
海　口 Haikou	3	0	9	0
三　亚 Sanya	0	0	0	0

8-13 沿海地带污染治理项目情况
Pollution Treatment Projects in Coastal Counties

单位：个 (unit)

地 区 Region	当年安排施工项目 Arranged for Construction in the Year		当年竣工项目 Completed in the Year	
	治理废水 Treatment of Waste Water	治理固体废物 Treatment of Solid Wastes	治理废水 Treament of Waste Water	治理固体废物 Treatment of Solid Wastes
天 津 Tianjin	**41**	**0**	**9**	**0**
河 北 Hebei	**1**	**0**	**2**	**0**
唐 山 Tangshan	1	0	2	0
秦皇岛 Qinhuangdao	0	0	0	0
沧 州 Cangzhou	0	0	0	0
辽 宁 Liaoning	**4**	**0**	**5**	**0**
大 连 Dalian	2	0	4	0
丹 东 Dandong	0	0	0	0
锦 州 Jinzhou	0	0	0	0
营 口 Yingkou	0	0	0	0
盘 锦 Panjin	0	0	0	0
葫芦岛 Huludao	2	0	1	0
上 海 Shanghai	**22**	**4**	**21**	**4**
江 苏 Jiangsu	**22**	**1**	**33**	**2**
南 通 Nantong	18	0	23	1
连云港 Lianyungang	3	1	8	1
盐 城 Yancheng	1	0	2	0

8-13 续表1 continued

地 区 Region	当年安排施工项目 Arranged for Construction in the Year		当年竣工项目 Completed in the Year	
	治理废水 Treatment of Waste Water	治理固体废物 Treatment of Solid Wastes	治理废水 Treament of Waste Water	治理固体废物 Treatment of Solid Wastes
浙 江 Zhejiang	**63**	**4**	**71**	**4**
杭 州 Hangzhou	1	0	2	0
宁 波 Ningbo	15	1	27	1
温 州 Wenzhou	8	0	6	0
嘉 兴 Jiaxing	11	1	10	1
绍 兴 Shaoxing	17	2	15	2
舟 山 Zhoushan	5	0	5	0
台 州 Taizhou	6	0	6	0
福 建 Fujian	**49**	**6**	**40**	**5**
福 州 Fuzhou	4	0	2	0
厦 门 Xiamen	11	0	15	0
莆 田 Putian	1	0	1	0
泉 州 Quanzhou	29	6	16	5
漳 州 Zhangzhou	3	0	4	0
宁 德 Ningde	1	0	2	0
山 东 Shandong	**38**	**10**	**45**	**10**
青 岛 Qingdao	2	1	3	1
东 营 Dongying	12	4	14	4
烟 台 Yantai	15	5	19	5
潍 坊 Weifang	9	0	7	0
威 海 Weihai	0	0	0	0
日 照 Rizhao	0	0	0	0
滨 州 Binzhou	0	0	2	0

8-13 续表2 continued

地　区 Region	当年安排施工项目 Arranged for Construction in the Year		当年竣工项目 Completed in the Year	
	治理废水 Treatment of Waste Water	治理固体废物 Treatment of Solid Wastes	治理废水 Treament of Waste Water	治理固体废物 Treatment of Solid Wastes
广　东　Guangdong	**53**	**0**	**70**	**0**
广　州　Guangzhou	9	0	16	0
深　圳　Shenzhen	4	0	4	0
珠　海　Zhuhai	3	0	2	0
汕　头　Shantou	2	0	7	0
江　门　Jiangmen	11	0	14	0
湛　江　Zhanjiang	8	0	8	0
茂　名　Maoming	2	0	2	0
惠　州　Huizhou	4	0	5	0
汕　尾　Shanwei	0	0	0	0
阳　江　Yangjiang	0	0	0	0
东　莞　Dongguan	0	0	0	0
中　山　Zhongshan	0	0	0	0
潮　州　Chaozhou	1	0	2	0
揭　阳　Jieyang	9	0	10	0
广　西　Guangxi	**6**	**1**	**9**	**1**
北　海　Beihai	3	1	7	1
防城港　Fangchenggang	2	0	1	0
钦　州　Qinzhou	1	0	1	0
海　南　Hainan	**3**	**0**	**3**	**0**
海　口　Haikou	3	0	3	0
三　亚　Sanya	0	0	0	0

8-14 沿海区域海洋类型自然保护区建设情况
Construction of Marine-Type Nature Reserves in Coastal Regions

地 区 Region	保护区数量（个） Number of Nature Reserves (unit)	按保护级别分（个） By Level of Protection (unit)		按保护类型分（个） By Type of Protection (unit)				保护区面积（平方千米） Area(km^2)
		国家级 National	地方级 Provincial	海洋和海岸生态系统 Marine and Coastal Ecosystem	海洋自然历史遗迹 Marine Natural and Historical Relics	海洋生物多样性 Marine Biodiversity	其他 Others	
合 计 Total	**135**	**33**	**102**	**56**	**24**	**52**	**3**	**49 035**
环渤海经济区 Round-the-Bohai Sea Economic Zone	39	11	28	23	3	13		16 499
长江三角洲经济区 Yangtze River Delta Economic Zone	11	5	6	2		6	3	2 356
海峡西岸经济区 Economic Zone on the West Side of the Taiwan Straits	10	3	7	5	2	3		692
珠江三角洲经济区 Zhujiang River Delta Economic Zone	52	7	45	23		29		4 031
环北部湾经济区 Round-the-Beibu Gulf Economic Zone	23	7	16	3	19	1		25 457

8-15 沿海地区海洋类型自然保护区建设情况
Construction of Marine-Type Nature Reserves

地　区 Region	保护区数量（个） Number of Nature Reserves (unit)	按保护级别分（个） By Level of Protection (unit)		按保护类型分（个） By Type of Protection (unit)				保护区面积（平方千米） Area (km^2)
		国家级 National	地方级 Provincial	海洋和海岸生态系统 Marine and Coastal Ecosystem	海洋自然历史遗迹 Marine Natural and Historical Relics	海洋生物多样性 Marine Biodiversity	其他 Others	
合　计 Total	**135**	**33**	**102**	**56**	**24**	**52**	**3**	**49 035**
天　津 Tianjin	1	1		1				359
河　北 Hebei	5	1	4	3	1	1		743
辽　宁 Liaoning	15	5	10	10	2	3		9 860
上　海 Shanghai	4	2	2	1			3	941
江　苏 Jiangsu	4	1	3			4		724
浙　江 Zhejiang	3	2	1	1		2		691
福　建 Fujian	10	3	7	5	2	3		692
山　东 Shandong	18	4	14	9		9		5 537
广　东 Guangdong	52	7	45	23		29		4 031
广　西 Guangxi	3	3		3				460
海　南 Hainan	20	4	16		19	1		24 997

8-16 全国海洋生态监控区基本情况
Basic Condition of the Marine Ecological Monitoring Areas Throughout the Country

生态监控区 Ecological Monitoring Area	所在地 Location	面积（平方千米）Area (km^2)	主要生态系统类型 Major Types of Ecosystem	多样性指数* Diversity Indices		
				浮游植物 Phyto-plankton	大型浮游动物 Macrozoo-plankton	底栖生物 Macrobendthos
双台子河口 Shungtaizi Estuary	辽宁省 Liaoning Province	3 000	河 口 Estuary			
锦州湾 Jinzhou Bay	辽宁省 Liaoning Province	650	海 湾 Bay			
滦河口-北戴河 Luanhe River Mouth-Beidaihe	河北省 Hebei Province	900	河 口 Estuary	2.06	1.57	1.58
渤海湾 Bohai Bay	天津市 Tianjin Municipality	3 000	海 湾 Bay	1.13	2.16	2.23
莱州湾 Laizhou Bay	山东省 Shandong Province	3 770	海 湾 Bay	2.54	1.86	3.04
黄河口 Yellow River Mouth	山东省 Shandong Province	2 600	河 口 Estuary	2.67	1.28	2.09
苏北浅滩 North Jiangsu Bank	江苏省 Jiangsu Province	15 400	滩涂湿地 Mudflat and Wetland	1.54	2.07	1.31
长江口 Yangtze River Mouth	上海市 Shanghai Municipality	13 668	河 口 Estuary	1.25	2.04	1.30
杭州湾 Hangzhou Bay	上海市 浙江省 Shanghai Municipality Zhejiang Province	5 000	海 湾 Bay	1.84	1.81	1.43
乐清湾 Yueqing Bay	浙江省 Zhejiang Province	464	海 湾 Bay	2.10	2.18	1.60

8-16 续表 continued

生态监控区 Ecological Monitoring Area	所在地 Location	面积（平方千米） Area (km^2)	主要生态系统类型 Major Types of Ecosystem	多样性指数* Diversity Indices		
				浮游植物 Phyto-plankton	大型浮游动物 Macrozoo-plankton	底栖生物 Macrobendthos
闽东沿岸 Coastal East Fujian	福建省 Fujian Province	5 063	海 湾 Bay	1.84	2.42	2.30
大亚湾 Daya Bay	广东省 Guangdong Province	1 200	海 湾 Bay	2.88	2.69	2.58
珠江口 Zhujiang River Mouth	广东省 Guangdong Province	3 980	河 口 Estuary	2.03	3.06	1.25
雷州半岛西南沿岸 Southwest Coast of Leizhou Peninsula	广东省 Guangdong Province	1 150	珊瑚礁 Coral Reef			
广西北海 Beihai, Guangxi	广西壮族自治区 Guangxi Zhuang Nationality Autonomous Region	120	珊瑚礁 红树林 海草床 Coral Reef Mangroves Seagrass Bed			
北仑河口 Beilun River Mouth	广西壮族自治区 Guangxi Zhuang Nationality Autonomous Region	150	红树林 Mangroves			
海南东海岸 East Coast of Hainan	海南省 Hainan Province	3 750	珊瑚礁 海草床 Coral Reef Seagrass Bed			
西沙珊瑚礁 Xisha Coral Reef	海南省 Hainan Province	400	珊瑚礁 Coral Reef			

注:*生物多样性指数：是生物种数和种类间个体数量分配均匀性的综合表现，用Shannon-Wiener多样性指数表征。

Note: Biodiversity index: refers to the comprehensive expression of the distributive homogeneity of the number of biological species and the number of individuals between varieties characterized by the Shannon-Wiener biodiversity index.

8-17 沿海地区风暴潮灾害情况
Survey of Storm Surges by Coastal Regions

受灾地区 Disaster Area	受灾人口 （万人） Disaster-stricken Population (10 000 persons)	死亡人数* （人） Death Toll (person)	受灾面积 Disaster-stricken area 农田 （千公顷） Disaster-Affected farmland (1 000 hm^2)	受灾面积 Disaster-stricken area 水产养殖 （千公顷） Affected Area of Mariculture (1 000 hm^2)
合　计 Total	**752.18**	**9**	**174.30**	**411.29**
天　津 Tianjin		0	0.00	0.17
河　北 Hebei	23.00	0	5.39	10.04
上　海 Shanghai	0.00	0	0.00	0.00
江　苏 Jiangsu	0.04	0		42.28
浙　江 Zhejiang		0		48.47
福　建 Fujian	1.42	0	4.81	5.78
山　东 Shandong	454.30	0	55.30	224.86
广　东 Guangdong	204.02	9	108.80	5.88
广　西 Guangxi	69.40	0		103.81

8-17 续表 continued

受灾地区 Disaster Area	海岸工程 （千米） Coastal Engineering (km)	房屋 House （间） (unit)	船只 （艘） Boats (unit)	直接经济损失 （亿元） Direct Economic Loss (100 million yuan)
合 计 Total	**332.38**	**36 006**	**2 338**	**126.29**
天 津 Tianjin	0.80	0	0	0.04
河 北 Hebei	4.15	16 000	152	20.44
上 海 Shanghai	0.30	0	0	0.06
江 苏 Jiangsu		233		6.15
浙 江 Zhejiang	259.39	0	915	42.57
福 建 Fujian	3.87	6	167	2.64
山 东 Shandong	36.65	17 658	597	31.59
广 东 Guangdong	6.15	1 391	506	17.47
广 西 Guangxi	21.07	718	1	5.33

注：*包括失踪人数。

Notes:*Includes lost people.

8-18 沿海地区赤潮灾害情况
Survey of Red Tide by Coastal Regions

时间 Date	影响区域 Area	最大面积（平方千米） Disaster Area(km^2)
累 计 **In the Aggregate**		**628**
其中：Including:		
5月23日～6月8日 May.23-Jun.8	浙江省温州南麂列岛附近海域 Sea Area Adjacent to the Nanji Archipelago, Wenzhou, Zhejiang Province	40
5月24日～6月3日 May.24-Jun.3	浙江省温州洞头岛附近海域 Sea area Adjacent to Dongtou Island, Wenzhou, Zhejiang Province	40
6月3日～6月7日 Jun.3-Jun.7	浙江省舟山嵊泗海域 Shengsi Sea Area of Zhoushan, Zhejiang Province	240
5月18日～6月7日 May.18-Jun.7	福建省宁德三沙湾的霞浦、福鼎海域 Xiapu and Fuding Sea Areas of Sansha Bay, Ningde, Fujian Province	130
5月27日～6月8日 May.27-Jun.8	福建省福州连江黄岐海域 Huangqi Sea Area of Lianjiang, Fuzhou, Fujian Province	40
5月30日～6月8日 May.30-Jun.8	福建省福州福清东翰海域 Donghan Sea Area of Fuqing, Fuzhou, Fujian Province	6
6月5日～6月8日 Jun.5-Jun.8	福建省福州罗源县碧里乡吉壁-新沃海域 Jibi and Xinwo Sea Areas of Bili Township Luoyuan County, Fuzhou, Fujian Province	10
5月26日～6月7日 May.26-Jun.7	福建省平潭海域 Pingtan Sea Area of Fujian Province	80
5月25日～5月27日 May.25-May.27	福建省莆田湄洲岛洋屿海域 Yangyu Sea Area of Meizhou Island, Putian, Fujian Province	2
5月30日～6月3日 May.30-Jun.3	福建省莆田东岙、坑口、古城、湄洲岛洋屿海域 Sea Area of Dong'an, Kengkou, Gucheng and Yangyu of Meizhou Island, Fujian Province	32
5月25日～5月26日 May.25-May.26	福建省泉州惠安杜厝海域 Ducuo Sea Area of Hui'an, Quanzhou, Fujian Province	1
5月30日～6月2日 May.30-Jun.2	福建省泉州惠安小岞海域 Xiaozuo Sea Area of Hui'an, Quanzhou, Fujian Province	7

主要统计指标解释

1. 工业废水排放量 指经过企业厂区所有排放口排到企业外部的工业废水量。包括生产废水、外排的直接冷却水、超标排放的矿井地下水和与工业废水混排的厂区生活污水,不包括外排的间接冷却水(清污不分流的间接冷却水应计算在内)。

2. 直接排入海的工业废水量 指经企业位于海边的排放口，直接排入海中的废水量。直接排入是指废水经过工厂的排污口直接排入海，而未经过城市下水道或其他中间体，也不受其他水体的影响。

3. 工业废水处理量 指报告期内各种水治理设施实际处理的工业废水量,包括处理后外排的和处理后回用的工业废水量，虽经处理但未达到国家或地方排放标准的废水量也应计算在内。计算时，如遇车间和厂排放口均有治理设施，并对同一废水分级处理时，不应重复计算工业废水处理量。

4. 工业废水处理率 指工业废水处理量占需要处理的工业废水量的百分率。其计算公式是:

工业废水处理率 =（工业废水处理量 / 需处理的工业废水量）×100%

式中，需处理工业废水量 =工业废水排放量+工业废水处理回用量-(工业废水排放达标量-工业废水处理排放达标量)

5. 工业废水排放达标量 指各项指标都达到国家或地方排放标准的外排工业废水量,包括经过处理后外排达标的和未经处理外排达标的两部分。国家排放标准见 GB 8978-88。

6. 工业废水处理排放达标量 指经过各种水治理设施处理后达到国家或地方排放标准的废水排放量。

7. 一般工业固体废物处置量 指将固体废物焚烧或者最终置于符合环境保护规定要求的场所并不再回取的工业固体废物量(包括当年处置往年的工业固体废物累计贮存量)。

8. 一般工业固体废物倾倒丢弃量 指将所产生的固体废物倾倒丢弃到固体废物污染防治设施、场所以外的量。不包括矿山开采的剥离废石和掘进废石(煤矸石和呈酸性或碱性的废石除外)。

9. 当年开工污染治理项目数 指报告期内由国家、部门、地方或企业单位安排开工的，并以治理废水、废气、固体废物、噪声和其他(如电磁波、恶臭等)环境污染的环境治理工程的总数。不包括“三同时”项目。

10. 当年竣工项目数 指报告期内竣工投入运行的治理废水、废气、固体废物、噪声及其他污染的环境工程项目的总数。

Explanatory Notes on Main Statistical Indicators

1. Volume of Industrial Waste Water Discharged refers to the quantity of industrial waste water discharged externally through all the outlets in the factory area of the enterprise, including the waste water from production, externally discharged direct cooling water, mine-shaft groundwater discharged exceeding the set standard and the domestic sewage of the factory area discharged together with the industrial waste water, but not including the indirect cooling water discharge externally.

2. Volume of Industrial Waste Water discharged Directly to Sea refers to the quantity of waste water directly discharged into the sea through the outlets of the enterprise by the sea. Direct discharge means the direct discharge into the sea of waste water through the outlets of the factory, which is not discharged via the urban sewers or other intermediates and is not affected by other water bodies.

3. Volume of Industrial Waste Water Treated refers to the industrial waste water volume actually treated by various water treatment facilities in the period covered by the report, including the quantity of the industrial waste water discharged and reused after treatment. The amount of the waste water which is not up to the state or local standard of discharge upon treatment should be included. If the workshops and outlets of the factory are provided with treatment facilities and carry out graded treatment of the same waste water, the processed volume of industrial waste water cannot be calculated repeatedly.

4. Rate of Industrial Waste Water Treatment refers to the percentage of the processed volume of industrial waste water in the industrial waste water volume that needs to be treated. The calculation formula is:

Rate of Industrial Waste Water Treatment = (Processed Volume of Industrial Waste Water/Processed Volume of Industrial Waste Water Required) × 100%.

Where: Processed Volume of Industrial Waste Water Required = (Processed Volume of Industrial Waste Water + Reused Volume of Industrial Waste Water After Treatment−Up-to-Standard Volume of Industrial Waste Water for Discharge−Up-to-Standard Volume of Industrial Waste Water for Discharge after Treatment).

5. Volume of Up-to-Standard Industrial Waste Water Discharged refers to the externally discharged volume of industrial waste water with all its indexes up to the state or local standard for discharge, including the volume of industrial waste water up to the standard for discharge after being treated or that without being treated. See GB8978–88 for the state′s discharge standard.

6. Volume of Up-to-Standard Discharged Industrial Waste Water After Treatment refers to the discharged volume of waste water that is up to the state or local standard for discharge upon treatment by various water treatment facilities.

7. Volume of Common Industrial Solid Wastes Discharged refers to the amount of solid wastes discharged out of the facilities and sites for the solid wastes pollution prevention and control, not including the stripped and tunneled waste ores in the excavation of mines (other than gangues and acid or alkaline waste ores).

8. Volume of Common Industrial Solid Wastes Treated refers to the volume of industrial solid wastes which are to be burned or finally placed at the sites in keeping with the requirement of environmental protection and will not be recovered (including the accumulated amount of storage in former years of industrial solid radioactive matter disposed of in the year).

9. Number of Pollution Treatment Projects Started in the Current Year refers to the total number of environmental pollution control projects for controlling waste water, waste gas, solid wastes, noise and other environmental pollutions (such as electromagnetic wave, offensive odor) started by the state, governmental departments, local governments or enterprises in the period covered by the report mainly for the purpose of pollution control and multipurpose utilization of ″three wastes″.

10. Number of Pollution Treatment Projects Completed in the Current Year refers to the total number of environmental engineering projects for controlling waste water, waste gas, solid wastes, noise and other pollutions which are completed and put into operation in the period covered by the report.

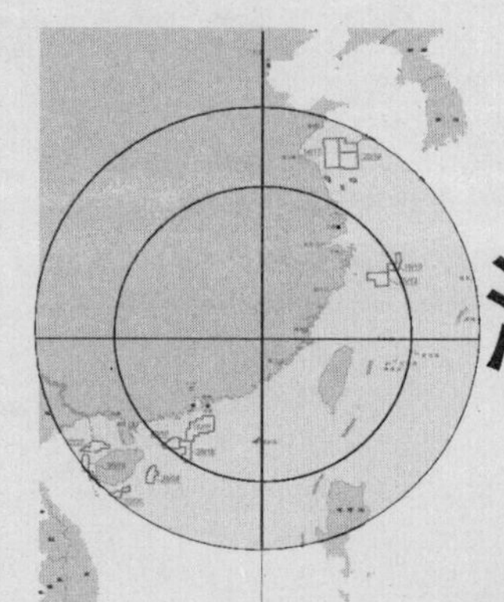

9 海洋行政管理及公益服务

Marine Administration and Public Service

9-1 海域使用管理情况
Sea Area Use Management

地 区 Region	发放海域使用权证书（本） Certificates of Right of Sea Area Use Issued (volume)	确权海域面积（公顷） Area of Waters with Established Rights (hm^2)	海域使用金（万元） Charge for Sea Area Utilization (10 000 yuan)
全国总计 National Total	**2 348**	**271 689.85**	**968 468.50**
天 津 Tianjin	36	1 084.81	159 243.72
河 北 Hebei	43	3 435.34	125 450.39
辽 宁 Liaoning	531	135 811.76	115 468.09
上 海 Shanghai	10	54.19	1 611.82
江 苏 Jiangsu	294	44 166.42	77 738.10
浙 江 Zhejiang	164	3 480.01	141 811.48
福 建 Fujian	140	4 185.74	65 807.76
山 东 Shandong	434	66 400.06	72 602.96
广 东 Guangdong	282	7 906.69	147 584.00
广 西 Guangxi	331	3 070.99	42 548.56
海 南 Hainan	83	2 093.84	9 050.13
其 他* other			9 551.49

注：*为沿海省（自治区、直辖市）管理海域以外（指渤海中部海域）。

Note: * Sea areas outside the control of coastal provinces of autonomous regions and municipalities directly under the Central Government. (Referring to the mid-Bohai Sea area)

9-2 海洋倾废管理情况
Management on the Ocean Dumping of Wastes

海 区 Sea Area	签发疏浚物海洋倾倒许可证（份） Permit Issued for Ocean Dumping of Dredged Materials (unit)	新选划倾倒区数（个） Number of Newly Designated Dumping Zones (unit)
合 计 **Total**	**146**	**18**
渤黄海 Bohai and Huanhai Sea	35	6
东 海 East China Sea	39	3
南 海 South China Sea	72	9

9-3 海洋执法检查情况
Marine Law Enforcement Inspection

执法领域 Law Enforcement Area	检查项目（个） Number of Items Subject to Inspection (unit)	检查次数（次） Number of Times of Inspection (time)	发现违法行为（起） Illegal Acts Found (case)
合　计 **Total**	**27 531**	**73 311**	**1 443**
渔业用海 Sea Use for Fishery	16 883	36 996	627
工业用海 Sea Use for Industrial Purposes	3 376	7 777	394
交通运输用海 Sea Use for Transportation and Communications	2 184	7 430	75
旅游娱乐用海 Sea Use for Tourism and Recreation	638	5 344	30
海底工程用海 Sea Use for Undersea Engineering	102	933	
排污倾倒用海 Sea Use for Sewage Discharge and Dumping	453	1 544	19
造地工程用海 Sea Use for Reclaiming Land from Sea	3 352	10 980	254
特殊用海 Special Sea Uses	137	668	16
其他用海 Other Sea Uses	406	1 639	28

9-4 沿海地区海滨观测台站分布概况
Distribution of Coastal Observation Stations by Coastal Regions

单位：个 (unit)

地 区 Region	合 计 Total	海洋站 Marine Station	验潮站① Tide Station	气象台站② Meteorological Station	地震台站 Seismic Station	雷达站 Radar Station
合 计 Total	**711**	**106**	**249**	**95**	**118**	**143**
天 津 Tianjin	14	1	0	0	9	4
河 北 Hebei	21	4	0	3	8	6
辽 宁 Liaoning	61	7	3	22	18	11
上 海 Shanghai	84	6	61	1	2	14
江 苏 Jiangsu	68	4	26	19	15	4
浙 江 Zhejiang	85	21	36	4	5	19
福 建 Fujian	85	15	11	25	13	21
山 东 Shandong	92	18	22	6	23	23
广 东 Guangdong	143	16	82	7	12	26
广 西 Guangxi	33	5	5	8	9	6
海 南 Hainan	25	9	3	0	4	9

注：①潮流量观测站43处，潮水位观测站206处。
②中国气象局所属的气象台站仅指沿海气象台站中有海洋气象观测、预报及服务等业务的台站，合计中含有中央气象台1个。

Notes: ①There are 43 tidal current observation stations and 206 tidal level observation stations.
② The meteorological observatories and stations under the China Meteorological Administration only refer to those with marine meteorotogical observation, forecast and service among the coastal meteorological observatories and the total includes the central meteorological station.

9-5 海洋预报服务概况
Marine Forecast Service

单位：次 (time)

预报项目 Item	数值预报 Numerical Forecast			
	预报服务次数 Frequency	发布次数 Frequency of Release		
		广播电视 Radio and TV	互联网 Internet	纸 质 Paper Media
合 计 Total	**36 202**	**365**	**35 343**	**494**
海 浪 Sea Wave	4 144		4 022	122
海 温 Sea Surface Temperature	2 318		2 196	122
潮 汐 Tide	2 557		2 557	
海 流 Sea Current	6 579	365	6 214	
海平面 Sea Level				
盐 度 Salinity	732		732	
赤 潮 Red Tide				
滨海旅游 Coastal Tourism				
海 冰 Sea Ice	183		183	
绿 潮 Green Seaweed	1 373		1 268	105
溢 油 Oil Spill	52		45	7
厄尔尼诺 EL Niño	9		2	7
专 项 Spccial Itcm	17 235		17 104	131
其 他 Others	1 020		1 020	

9-5 续表 continued

预报项目 Item	统计预报 Statistical Forecast			
	预报服务次数 Frequency	发布次数 Frequency of Release		
		广播电视 Radio and TV	互联网 Internet	纸 质 Paper Media
合 计 Total	**192 281**	**56 571**	**75 816**	**59 894**
海 浪 Sea Wave	56 531	18 747	23 388	14 396
海 温 Sea Surface Temperature	47 672	17 346	20 143	10 183
潮 汐 Tide	42 990	15 688	17 472	9 830
海 流 Sea Current				
海平面 Sea Level				
盐 度 Salinity	575	184	184	207
赤 潮 Red Tide	1 064	202	294	568
滨海旅游 Coastal Tourism	10 690	3 413	4 662	2 615
海 冰 Sea Ice	2 291	154	126	2 011
绿 潮 Green Seaweed				
溢 油 Oil Spill	58			58
厄尔尼诺 EL Niño	9		2	7
专 项 Special Item	26 403	472	9 118	16 813
其 他 Others	3 998	365	427	3 206

注：本表只包含国家海洋局资料。

Note: This table only contains the data from the State Oceanic Administration.

9-6 海洋环境观测情况
Condition of Marine Environmental Observation

项目 Item	台站观测 Station Observation	断面观测 Sectional Observation	浮标观测 Buoy Observation	船舶测报 Ship Measuring and Reporting	其他观测 Other Observation
测站（点）数（个） Number of Stations (unit)	108	119	41	57	154
实际获得数据量（个） Quantity of Data Actually Obtained (unit)	217 532 839	20 767	3 699 426	11 716 945	51 785 907

9-7 海洋调查概况
Marine Survey Statistics

调查名称 Name	站点数（个） Number of Stations (unit)	船舶数（艘） Number of Ships (unit)	项目数（个） Number of Items (unit)	实际获得数据（个） Quantity of Data Actually Obtained (unit)	发布通（公、简）报量（期） Quantity of Circulars (Bulletin, Brief Reports) (unit)
合　计 Total	**4 279**	**347**	**427**	**225 820**	**110**
大洋调查 Oceanic Survey	1 561	5	12	39 710	13
极地调查 Polar Survey	152	2	17	6 486	
专项调查 Special Survey	1 529	34	63	80 686	23
其他调查 Other Surveys	1 037	306	335	98 938	74

9-8 涉外海洋科学研究审批情况
Examination and Approval of Foreign-Related Marine Scientific Research

单位：份 (unit)

审批单位 Examing and Approving Units	审批研究活动申请 Application for Examing and Approving Research Activities	船只作业计划审批 Examination and Approval of the Ship Operating Plan
国家海洋局 State Oceanic Administration , People's Republic of China	4	4

9-9 海洋档案及利用情况
Marine File and Its Use

指　标 Item	指 标 值 Data
室（馆）存档案 Files Deposited in the Archives	
纸介质档案（卷、册） Paper Media Files (reel,volume)	111 183
电子档案（GB） Electronic Media Files (GB)	51 096
本年接收档案 Files received this year	
纸质档案（卷、册） Paper Media Files (reel,volume)	18 214
电子档案（GB） Electronic Media Files (GB)	9 898
本年利用档案 Files utilized this year	
纸质档案（件） Paper Media Files (piece)	13 647
电子档案（GB） Electronic Media Files (GB)	5 623

9-10 卫星遥感接收应用情况
Remote-Sensing Receiving and Utilization

指　标 Item	指标值 Data
卫星接收次数（轨） Number of Satellite Receptions (orbit / track)	97 168
全年接收时间（天） Whole Year Receiving Time (day)	7 754
全年实际接收存档数据量（GB） Amount of Data Actually Received and Placed on File in the Whole Year (GB)	31 648.50
累计存档数据量（GB） Total Amount of Data Placed on File (GB)	67 078.36
卫星数据分发 Satellite Data Distribution	
类别用户（个） Classified Users (unit)	84
分发数据量（GB） Amount of Data Distributed (GB)	58 808.03

9-11 海洋标准化监督管理情况
Supervision and Management of Marine Standardization

单位：项　　　　(item)

指　标 Item	指标值 Data
标准立项审查 Examination of the Standards for Authorization	66
国家标准 National Standards	13
行业标准 Professional Standards	53
标准审查 Examination of Standards	97
国家标准 National Standards	33
行业标准 Professional Standards	64
标准出版 Standards Publication	1
国家标准 National Standards	
行业标准 Professional Standards	1
标准实施监督检查（次） Supervision and Examination of Standards Implementation (time)	13

主要统计指标解释

1. 海域使用检查 针对不同类型的用海行为进行的监督检查。

2. 涉外海洋科研项目检查 主要针对国际组织、外国组织和个人为和平目的，单独或者与中华人民共和国的组织合作，使用船舶或者其他运载工具、设施，在中华人民共和国内海、领海以及中华人民共和国管辖的其他海域内进行的对海洋环境和海洋资源等的调查研究活动进行监督检查。

3. 海底电缆管道检查 主要是针对铺设海底电缆管道路由调查、铺设施工和维修改造等的监督检查。

4. 海洋工程建设项目环境保护检查 主要是针对防治海洋工程建设项目对海洋环境的污染损害的监督检查。

5. 海洋倾废检查 主要是针对防治倾倒废弃物对海洋环境的污染损害的监督检查。

6. 海洋生态保护检查 主要是针对红树林、珊瑚礁、滨海湿地、海岛、海湾、入海河口、重要渔业水域等具有典型性、代表性的海洋生态系统，珍稀、濒危海洋生物的天然集中分布区，具有重要经济价值的海洋生物生存区域及有重大科学文化价值的海洋自然历史遗迹和自然景观等海洋自然保护区以及其他需要予以特殊保护的区域的监督检查。

7. 断面监测 按照国家海洋局“断面监测方案”，每年定期利用船舶在沿海设定的断面上进行海洋水文、气象、生物、化学等项目的监测活动。

8. 浮标监测 在海上固定站位获取长期、连续海洋环境观测资料的海上锚定资料浮标。

9. 船舶监测 利用在固定航线上走航的商船或渔船，每日定时所在地的海洋环境状况（主要是水文气象要素），并将监测数据实时发送给有关单位，供海洋环境预报使用。

10. 大洋调查 以大洋科考、研究为目的的远洋调查。

11. 专项调查 为完成国家专项任务进行的海洋调查。

12. 全年接收时间 指全年整个应用系统接收的各类遥感卫星数据时，接收设备工作时间。

13. 全年实际接收存档数据量 指全年整个应用系统接收的各类遥感卫星数据原始数据量。

14. 累计存档数据量 指全年整个应用系统制作的各类遥感产品数据量。

15. 分发数据量 指应用系统提供数据产品用于开展各类应用的数据量。

16. 卫星接收次数 地面观测系统接收卫星数据的数量。

17. 国家标准 针对海洋领域内需要在全国范围内统一的有关技术要求所制定的国家标准。海洋国家标准由国家标准化主管部门统一批准、编号和发布。

18. 行业标准 对没有海洋国家标准而又需要在海洋领域内统一的技术要求所制定的标准。海洋行业标准由国家海洋局统一批准、编号和发布。

Explanatory Notes on Main Statistical Indicators

1. Sea Area Use Supervision and inspection refers to the various types of sea area use conducts.

2. Inspection of Foreign-Related Marine Scientific Research Projects refers to the supervision and inspection of the activities of surveying marine environment and resources conducted by international organizations, foreign organizations and individuals for peaceful purposes, alone or in cooperation with PRC organizations, by using ships or other means of delivery as facilities in the internal seas and territorial waters of the People's Republic of China as well as in the other water under the jurisdiction of the People's Republic of China.

3. Inspection of Submarine Cables and Pipelines is mainly aimed at the survey of the submarine cables and pipelines laying route and supervision and inspection of the laying construction, repair and transformation.

4. Environmental Protection Inspection for the Marine Engineering Construction Projects is mainly directed against preventing and controlling the pollution damage of the marine engineering construction project to the marine environment.

5. Inspection of Oceanic Dumping of Wastes mainly refers to the supervision and inspection aimed at preventing and controlling the pollution damage of wastes dumping to the marine environment.

6. Inspection of Marine Ecological Protection mainly refers to the supervision and inspection of the typical and representative marine ecosystems such as mangroves, coral reef, coastal wetland, sea island, bay, estuaries open to the sea, important fishery waters, etc, the natural centralized distribution zones of rare and endangered marine life, the living areas for the marine life with significant economic values and the marine nature reserves such as the marine natural and historical remains and natural landscapes with important scientific and cultural values as well as other areas needing to be specially protected.

7. Sectional Monitoring According to the *"Sectional Monitoring Plan"* of the State Oceanic Administration, monitoring activities concerning such items as marine hydrology, meteorology, biology and chemistry are carried out regularly every year on the sections set in the coastal area.

8. Buoy Monitoring refers to the monitoring carried out by the offshore mooring data buoys which acquire long-term, continuous marine environmental observations at the fixed stations at sea.

9. Ship Monitoring Monitoring the marine environmental condition in the sea area of fixed times every day by using the merchant or fishing vessels cruising on the fixed navigation line (mainly the hydro meteorological elements) and transmitting the monitored data in real time to the related units for use in the marine environmental forecast.

10. Oceanic Survey refers to the oceanic surveys aimed at the oceanic scientific investigations and research.

11. Special-Subject Survey refers to the oceanic investigation for the purpose of fulfilling the state′s special tasks

12. Whole Year Receiving Time refers to the operating time of receiving equipment in the whole year when the whole application system receives all types of remote sensing satellite data.

13. Amount of Data Actually Received and Placed on File Throughout the year refers to the raw data of remote-sensing satellite data of various kinds received by the whole application system throughout the year.

14. Total Amount of Date Placed on File refers to the date amount of various remote-sensing products made by the whole application systems throughout the year.

15. Amount of Data Distributed refers to the amount of data products provided by the application system for various uses.

16. Number of Satellite Receptions refers to the amount of data received by the ground observation system.

17. National Standards refers to the standards formulated in view of the relevant technical requirements in the marine field that need to be unified throughout the country. The Marine National standards are approved, numbered and issued uniformly by the state department responsible for standardization.

18. Professional Standards refers to the standards formulated for the technical requirements which have no national standards but need to be unified in the marine field. The Marine Professional Standards are approved, numbered and issued by the State Oceanic Administration.

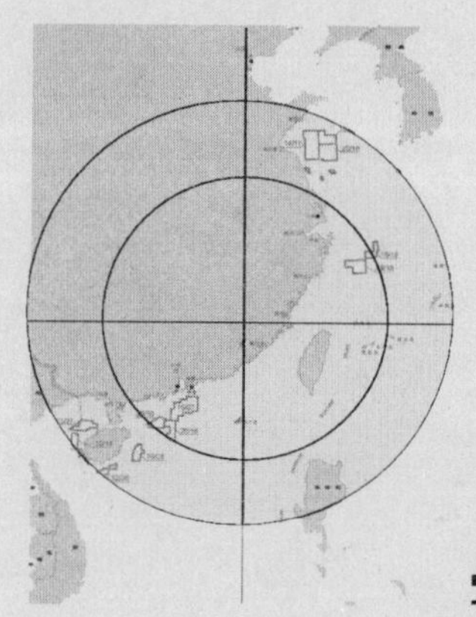

10

全国及沿海社会经济

National and Coastal Socioeconomy

10-1 国内生产总值
Gross Domestic Product

单位：亿元 (100 million yuan)

年 份 Year	国内生产总值 Gross Domestic Product	第一产业 Primary Industry	第二产业 Secondary Industry	第三产业 Tertiary Industry
2001	109 655.2	15 781.3	49 512.3	44 361.6
2002	120 332.7	16 537.0	53 896.8	49 898.9
2003	135 822.8	17 381.7	62 436.3	56 004.7
2004	159 878.3	21 412.7	73 904.3	64 561.3
2005	184 937.4	22 420.0	87 598.1	74 919.3
2006	216 314.4	24 040.0	103 719.5	88 554.9
2007	265 810.3	28 627.0	125 831.4	111 351.9
2008	314 045.4	33 702.0	149 003.4	131 340.0
2009	340 902.8	35 226.0	157 638.8	148 038.0
2010	401 512.8	40 533.6	187 383.2	173 596.0
2011	473 104.0	47 486.2	220 412.8	205 205.0
2012	518 942.1	52 373.6	235 162.0	231 406.5

10-2 国内生产总值增长速度
Growth Rate of Gross Domestic Product

单位：% (%)

年 份 Year	国内生产总值 Gross Domestic Product	第一产业 Primary Industry	第二产业 Secondary Industry	第三产业 Tertiary Industry
2001	8.3	2.8	8.4	10.3
2002	9.1	2.9	9.8	10.4
2003	10.0	2.5	12.7	9.5
2004	10.1	6.3	11.1	10.1
2005	11.3	5.2	12.1	12.2
2006	12.7	5.0	13.4	14.1
2007	14.2	3.7	15.1	16.0
2008	9.6	5.4	9.9	10.4
2009	9.2	4.2	9.9	9.6
2010	10.4	4.3	12.3	9.8
2011	9.3	4.3	10.3	9.4
2012	7.7	4.5	7.9	8.1

注：本表按可比价格计算(上年为基期)。

Note:This table is calculated at the comparable price (with the previous year as the base period).

10-3 沿海地区生产总值
Gross Regional Product of Coastal Regions

单位：亿元 (100 million yuan)

地 区 Region	地区生产总值 Gross Regional Product	第一产业 Primary Industry	第二产业 Secondary Industry	第三产业 Tertiary Industry
合 计 Total	**315 894.17**	**22 517.63**	**156 867.46**	**136 509.08**
天 津 Tianjin	12 893.88	171.60	6 663.82	6 058.46
河 北 Hebei	26 575.01	3 186.66	14 003.57	9 384.78
辽 宁 Liaoning	24 846.43	2 155.82	13 230.49	9 460.12
上 海 Shanghai	20 181.72	127.80	7 854.77	12 199.15
江 苏 Jiangsu	54 058.22	3 418.29	27 121.95	23 517.98
浙 江 Zhejiang	34 665.33	1 667.88	17 316.32	15 681.13
福 建 Fujian	19 701.78	1 776.71	10 187.94	7 737.13
山 东 Shandong	50 013.24	4 281.70	25 735.73	19 995.81
广 东 Guangdong	57 067.92	2 847.26	27 700.97	26 519.69
广 西 Guangxi	13 035.10	2 172.37	6 247.43	4 615.30
海 南 Hainan	2 855.54	711.54	804.47	1 339.53

10-4 沿海地区生产总值增长速度
Growth Rate of Gross Regional Product of Coastal Regions

单位：%　　　　(%)

地　区 Region	2002	2003	2004	2005	2006	2007	2008	2009	2010	2011	2012
天　津 Tianjin	12.7	14.8	15.8	14.9	14.7	15.5	16.5	16.5	17.4	16.4	13.8
河　北 Hebei	9.6	11.6	12.9	13.4	13.4	12.8	10.1	10.0	12.2	11.3	9.6
辽　宁 Liaoning	10.2	11.5	12.8	12.7	14.2	15.0	13.4	13.1	14.2	12.2	9.5
上　海 Shanghai	11.3	12.3	14.2	11.4	12.7	15.2	9.7	8.2	10.3	8.2	7.5
江　苏 Jiangsu	11.7	13.6	14.8	14.5	14.9	14.9	12.7	12.4	12.7	11.0	10.1
浙　江 Zhejiang	12.6	14.7	14.5	12.8	13.9	14.7	10.1	8.9	11.9	9.0	8.0
福　建 Fujian	10.2	11.5	11.8	11.6	14.8	15.2	13.0	12.3	13.9	12.3	11.4
山　东 Shandong	11.7	13.4	15.4	15.0	14.7	14.2	12.0	12.2	12.3	10.9	9.8
广　东 Guangdong	12.4	14.8	14.8	14.1	14.8	14.9	10.4	9.7	12.4	10.0	8.2
广　西 Guangxi	10.6	10.2	11.8	13.1	13.6	15.1	12.8	13.9	14.2	12.3	11.3
海　南 Hainan	9.6	10.6	10.7	10.5	13.2	15.8	10.3	11.7	16.0	12.0	9.1

注：本表按可比价格计算(上年为基期)。

Note: This table is calculated at the comparable price (with the previous year as the base period).

10-5 沿海城市生产总值（2011年）
Gross Regional Product of Coastal Cities, 2011

单位：亿元 (100 million yuan)

沿海城市 Coastal City		地区生产总值 Gross Regional Product	第一产业 Primary Industry	第二产业 Secondary Industry	第三产业 Tertiary Industry
合　计	**Total**	**168 908.86**	**9 069.12**	**83 979.67**	**75 860.17**
天　津	**Tianjin**	**11 307.28**	**159.72**	**5 928.32**	**5 219.24**
河　北	**Hebei**	**9 097.73**	**922.31**	**5 048.11**	**3 127.32**
唐　山	Tangshan	5 442.45	486.53	3 269.93	1 686.00
秦皇岛	Qinhuangdao	1 070.08	139.94	419.48	510.66
沧　州	Cangzhou	2 585.20	295.84	1 358.70	930.66
辽　宁	**Liaoning**	**11 150.86**	**958.41**	**5 963.33**	**4 229.11**
大　连	Dalian	6 150.63	395.72	3 204.18	2 550.73
丹　东	Dandong	888.67	119.70	457.15	311.82
锦　州	Jinzhou	1 116.93	173.37	554.01	389.54
营　口	Yingkou	1 224.65	90.76	673.81	460.09
盘　锦	Panjin	1 119.92	94.33	762.17	263.41
葫芦岛	Huludao	650.06	84.53	312.01	253.52
上　海	**Shanghai**	**19 195.69**	**124.94**	**7 927.89**	**11 142.86**
江　苏	**Jiangsu**	**8 262.07**	**908.15**	**4 182.02**	**3 171.90**
南　通	Nantong	4 080.22	287.21	2 221.48	1 571.53
连云港	Lianyungang	1 410.52	204.11	654.28	552.13
盐　城	Yancheng	2 771.33	416.83	1 306.26	1 048.24
浙　江	**Zhejiang**	**26 033.08**	**1 180.07**	**13 545.52**	**11 307.52**
杭　州	Hangzhou	7 019.06	236.77	3 323.79	3 458.50
宁　波	Ningbo	6 059.24	255.23	3 349.53	2 454.49
温　州	Wenzhou	3 418.53	107.88	1 760.72	1 549.94
嘉　兴	Jiaxing	2 677.09	142.80	1 537.26	997.03
绍　兴	Shaoxing	3 332.00	172.10	1 834.34	1 325.56
舟　山	Zhoushan	772.75	76.04	349.16	347.55
台　州	Taizhou	2 754.41	189.25	1 390.72	1 174.45
福　建	**Fujian**	**14 295.52**	**1 062.30**	**7 556.00**	**5 677.23**
福　州	Fuzhou	3 736.38	325.09	1 711.19	1 700.10
厦　门	Xiamen	2 539.31	24.68	1 297.15	1 217.49
莆　田	Putian	1 050.62	98.80	613.28	338.54
泉　州	Quanzhou	4 270.89	151.78	2 662.37	1 456.74
漳　州	Zhangzhou	1 768.20	293.30	836.26	638.64
宁　德	Ningde	930.12	168.65	435.75	325.72

10-5 续表 continued

沿海城市 Coastal City		地区生产总值 Gross Regional Product	第一产业 Primary Industry	第二产业 Secondary Industry	第三产业 Tertiary Industry
山　东	**Shandong**	**22 883.22**	**1 587.60**	**12 630.12**	**8 665.50**
青　岛	Qingdao	6 615.60	306.38	3 150.72	3 158.50
东　营	Dongying	2 676.35	99.18	1 914.81	662.36
烟　台	Yantai	4 906.83	361.43	2 830.88	1 714.52
潍　坊	Weifang	3 541.84	359.28	1 961.40	1 221.16
威　海	Weihai	2 110.95	171.18	1 139.36	800.41
日　照	Rizhao	1 214.07	112.08	660.66	441.33
滨　州	Binzhou	1 817.58	178.07	972.29	667.22
广　东	**Guangdong**	**44 097.94**	**1 741.01**	**20 236.86**	**22 120.12**
广　州	Guangzhou	12 423.44	204.54	4 576.98	7 641.92
深　圳	Shenzhen	11 505.53	6.55	5 343.32	6 155.65
珠　海	Zhuhai	1 404.93	36.55	764.41	603.98
汕　头	Shantou	1 275.74	73.76	649.70	552.28
江　门	Jiangmen	1 830.64	138.36	997.27	695.01
湛　江	Zhanjiang	1 700.23	344.15	710.43	645.65
茂　名	Maoming	1 745.31	319.27	696.66	729.38
惠　州	Huizhou	2 093.08	116.51	1 223.25	753.32
汕　尾	Shanwei	550.55	89.99	258.52	202.05
阳　江	Yangjiang	766.82	160.33	338.98	267.52
东　莞	Dongguan	4 735.39	17.88	2 366.20	2 351.32
中　山	Zhongshan	2 193.20	58.45	1 223.25	911.51
潮　州	Chaozhou	647.22	45.97	352.93	248.32
揭　阳	Jieyang	1 225.86	128.70	734.96	362.21
广　西	**Guangxi**	**1 557.00**	**329.25**	**715.75**	**512.01**
北　海	Beihai	496.58	115.45	207.39	173.75
防城港	Fangchenggang	413.77	57.79	217.63	138.35
钦　州	Qinzhou	646.65	156.01	290.73	199.91
海　南	**Hainan**	**1 028.47**	**95.36**	**245.75**	**687.36**
海　口	Haikou	733.91	53.76	182.05	498.10
三　亚	Sanya	294.56	41.60	63.70	189.26

注：本表各省数据为合计数。

Note: The data for the provinces are the totals.

10-6 沿海县主要统计指标（2011年）
Main Statistical Indicators of Coastal Counties, 2011

单位：万元 (10 000 yuan)

沿海县 Coastal County		第一产业增加值 Value-added of Primary Industry	第二产业增加值 Value-added of Secondary Industry	地方财政一般预算收入 General Budget Revene of Local Governments	地方财政一般预算支出 General Budget Expenditure of Local Governments
合　计	**Total**	**46 578 130**	**191 839 107**	**22 140 182**	**34 950 716**
河　北	**Hebei**	**2 757 477**	**4 744 761**	**411 648**	**1 089 947**
丰　南	Fengnan				
滦　南	Luannan	673 346	1 084 617	78 909	201 916
乐　亭	Leting	648 522	1 095 255	71 678	196 870
唐　海	Tanghai	162 667	347 138	60 000	131 033
昌　黎	Changli	556 419	658 342	50 284	158 822
抚　宁	Funing	413 346	589 269	64 437	150 279
黄　骅	Huanghua	238 139	853 805	71 890	178 163
海　兴	Haixing	65 038	116 335	14 450	72 864
辽　宁	**Liaoning**	**5 835 636**	**20 699 841**	**2 388 731**	**3 330 822**
长　海	Changhai	409 364	74 019	33 001	57 495
瓦房店	Wafangdian	804 902	5 391 413	510 988	597 656
普兰店	Pulandian	868 668	3 795 919	322 657	393 686
庄　河	Zhuanghe	1 010 039	3 953 751	370 227	461 621
东　港	Donggang	586 961	2 007 190	225 815	298 284
凌　海	Linghai	451 187	1 282 945	170 022	242 553
盖　州	Gaizhou	311 847	1 003 625	140 314	274 905
大　洼	Dawa	527 127	1 495 380	271 529	369 888
盘　山	Panshan	382 280	802 706	148 871	222 299
绥　中	Suizhong	324 617	465 942	110 218	234 145
兴　城	Xingcheng	158 644	426 951	85 089	178 290

10-6 续表1 continued

沿海县 Coastal County		第一产业增加值 Value-added of Primary Industry	第二产业增加值 Value-added of Secondary Industry	地方财政一般预算收入 General Budget Revene of Local Governments	地方财政一般预算支出 General Budget Expenditure of Local Governments
江　苏	**Jiangsu**	**6 418 661**	**21 795 800**	**3 542 338**	**5 435 445**
海　安	Hai'an	429 429	2 280 100	285 060	430 909
如　东	Rudong	504 966	2 231 300	253 906	424 067
启　东	Qidong	559 581	2 771 100	440 716	532 719
海　门	Haimen	413 915	3 482 200	423 512	476 495
赣　榆	Ganyu	546 920	1 444 400	238 099	446 294
东　海	Donghai	472 743	1 164 400	235 008	433 497
灌　云	Guanyun	442 922	902 900	216 093	359 118
灌　南	Guannan	338 538	913 800	222 193	376 717
响　水	Xiangshui	325 010	796 200	170 057	262 384
滨　海	Binhai	454 129	1 043 400	207 728	383 562
射　阳	Sheyang	636 574	1 177 800	190 100	358 292
东　台	Dongtai	714 934	2 054 300	359 200	537 016
大　丰	Dafeng	579 000	1 533 900	300 666	414 375
浙　江	**Zhejiang**	**5 702 684**	**47 701 938**	**6 117 421**	**7 700 943**
象　山	Xiangshan	530 238	1 491 559	248 303	395 453
宁　海	Ninghai	344 888	1 831 916	265 948	363 142
余　姚	Yuyao	403 795	3 942 933	550 163	582 540
慈　溪	Cixi	435 708	5 268 633	715 203	806 381
奉　化	Fenghua	260 258	1 260 280	219 607	329 670
洞　头	Dongtou	37 425	157 805	30 706	112 627
平　阳	Pingyang	127 304	1 158 903	167 528	288 683
苍　南	Cangnan	237 814	1 428 614	174 301	351 397
瑞　安	Rui'an	179 492	2 633 988	386 789	431 542
乐　清	Yueqing	184 035	3 466 154	416 600	457 727
海　盐	Haiyan	203 631	1 702 289	171 825	207 050
海　宁	Haining	242 622	3 207 716	388 194	398 663
平　湖	Pinghu	178 732	2 489 795	324 423	329 260

10-6 续表2 continued

沿海县 Coastal County		第一产业增加值 Value-added of Primary Industry	第二产业增加值 Value-added of Secondary Industry	地方财政一般预算收入 General Budget Revene of Local Governments	地方财政一般预算支出 General Budget Expenditure of Local Governments
绍 兴	Shaoxing	332 052	5 548 751	637 690	586 443
上 虞	Shangyu	365 586	2 964 379	355 195	357 568
岱 山	Daishan	216 000	871 485	84 685	245 605
嵊 泗	Shengsi	143 220	77 585	43 738	163 632
玉 环	Yuhuan	242 524	2 236 333	216 856	261 520
三 门	Sanmen	194 134	548 771	96 022	186 879
温 岭	Wenling	505 873	3 468 910	361 688	451 736
临 海	Linhai	337 353	1 945 139	261 957	393 425
福 建	**Fujian**	**6 000 509**	**31 096 689**	**2 987 913**	**4 778 380**
连 江	Lianjiang	773 347	920 400	146 923	226 435
罗 源	Luoyuan	218 974	936 500	61 104	95 563
平 潭	Pingtan	301 160	317 700	82 300	444 766
福 清	Fuqing	755 545	2 806 500	309 111	397 184
长 乐	Changle	335 246	2 553 100	186 922	255 220
仙 游	Xianyou	239 979	812 756	92 532	214 579
惠 安	Hui'an	209 504	1 990 846	178 106	239 131
石 狮	Shishi	169 321	2 481 254	240 012	302 586
晋 江	Jinjiang	174 033	7 390 136	639 164	701 225
南 安	Nan'an	215 912	3 722 400	284 426	402 827
云 霄	Yunxiao	206 360	364 630	30 918	102 506
漳 浦	Zhangpu	495 733	656 539	96 648	211 087
诏 安	Zhao'an	299 353	472 668	38 922	127 336
东 山	Dongshan	264 144	429 154	65 000	151 484
龙 海	Longhai	493 073	2 513 306	265 932	353 944
霞 浦	Xiapu	314 335	368 700	45 602	150 475
福 安	Fu'an	291 610	1 385 300	121 355	213 796
福 鼎	Fuding	242 880	974 800	102 936	188 236

10-6 续表3 continued

沿海县 Coastal County		第一产业增加值 Value-added of Primary Industry	第二产业增加值 Value-added of Secondary Industry	地方财政一般预算收入 General Budget Revene of Local Governments	地方财政一般预算支出 General Budget Expenditure of Local Governments
山　东	**Shandong**	**7 880 747**	**49 195 518**	**4 620 063**	**6 255 282**
胶　州	Jiaozhou	449 490	3 771 200	343 026	460 638
即　墨	Jimo	535 225	3 734 300	361 000	483 900
胶　南	Jiaonan	488 633	3 818 600	363 567	462 734
垦　利	Kenli	156 450	1 630 842	120 188	193 108
利　津	Lijin	240 013	942 974	61 336	139 569
广　饶	Guangrao	354 465	3 858 765	221 186	306 313
长　岛	Changdao	330 694	49 165	12 118	42 573
龙　口	Longkou	287 319	4 828 931	475 668	526 337
莱　阳	Laiyang	368 180	1 980 950	93 536	172 267
莱　州	Laizhou	550 135	3 038 614	303 737	388 687
蓬　莱	Penglai	244 809	2 260 496	200 675	255 498
招　远	Zhaoyuan	300 332	2 999 950	285 300	344 468
海　阳	Haiyang	482 459	1 199 906	162 008	246 639
寿　光	Shouguang	726 957	2 805 705	416 006	440 376
昌　邑	Changyi	294 178	1 585 878	158 136	221 078
文　登	Wendeng	479 784	3 453 753	342 329	393 274
荣　成	Rongcheng	667 736	3 698 045	360 589	579 736
乳　山	Rushan	296 044	1 884 540	172 678	264 063
无　棣	Wudi	325 966	1 136 740	96 872	178 504
沾　化	Zhanhua	301 878	516 164	70 108	155 520
广　东	**Guangdong**	**6 845 822**	**12 997 574**	**1 094 737**	**3 136 221**
南　澳	Nan'ao	32 039	47 490	12 506	57 262
台　山	Taishan	417 970	1 747 010	155 605	250 305
恩　平	Enping	158 737	422 700	59 668	151 650
遂　溪	Suixi	737 879	557 261	43 107	163 094
徐　闻	Xuwen	496 488	133 459	27 826	125 298

10-6 续表4 continued

沿海县 Coastal County		第一产业增加值 Value-added of Primary Industry	第二产业增加值 Value-added of Secondary Industry	地方财政一般预算收入 General Budget Revene of Local Governments	地方财政一般预算支出 General Budget Expenditure of Local Governments
廉　江	Lianjiang	711 187	825 297	56 065	227 329
雷　州	Leizhou	704 454	248 244	45 018	198 389
吴　川	Wuchuan	216 630	520 385	41 228	141 159
电　白	Dianbai	659 787	929 898	80 778	242 572
惠　东	Huidong	319 879	1 516 157	134 282	280 719
海　丰	Haifeng	277 818	793 873	90 038	179 445
陆　丰	Lufeng	386 839	695 858	97 010	253 493
阳　西	Yangxi	399 631	380 868	28 641	126 737
阳　东	Yangdong	320 313	855 131	71 678	151 261
饶　平	Raoping	285 063	694 896	34 323	183 733
揭　东	Jiedong	334 228	1 809 930	80 587	210 319
惠　来	Huilai	386 880	819 117	36 377	193 456
广　西	**Guangxi**	**690 977**	**660 437**	**97 300**	**369 449**
合　浦	Hepu	593 579	459 597	41 328	254 839
东　兴	Dongxing	97 398	200 840	55 972	114 610
海　南	**Hainan**	**4 445 617**	**2 946 549**	**880 031**	**2 854 227**
琼　海	Qionghai	567 147	211 370	126 265	304 820
儋　州	Danzhou	837 743	207 910	71 437	376 197
文　昌	Wenchang	591 598	322 587	85 278	304 715
万　宁	Wanning	323 444	306 040	82 185	275 873
东　方	Dongfang	250 767	498 378	62 235	229 457
澄　迈	Chengmai	376 425	692 239	125 833	490 903
临　高	Lingao	548 832	73 840	30 515	176 868
昌　江	Changjiang	176 621	423 005	71 286	165 755
乐　东	Ledong	352 958	72 734	57 371	240 823
陵　水	Lingshui	420 082	138 446	167 626	288 816

注：本表各省数据为合计数。

Note: The data for the provinces are the totals.

10-7 沿海地区财政收支
Financial Revenue and Expenditure by Coastal Regions

单位：亿元 (100 million yuan)

地 区 Region	公共财政预算收入 Public Budgetary Revenue	公共财政预算支出 Public Budgetary Revenue
合 计 **Total**	**33 635.59**	**45 951.59**
天 津 Tianjin	1 760.02	2 143.21
河 北 Hebei	2 084.28	4 079.44
辽 宁 Liaoning	3 105.38	4 558.59
上 海 Shanghai	3 743.71	4 184.02
江 苏 Jiangsu	5 860.69	7 027.67
浙 江 Zhejiang	3 441.23	4 161.88
福 建 Fujian	1 776.17	2 607.50
山 东 Shandong	4 059.43	5 904.52
广 东 Guangdong	6 229.18	7 387.86
广 西 Guangxi	1 166.06	2 985.23
海 南 Hainan	409.44	911.67

10-8 沿海地区教育基本情况
Basic Conditions of Education by Coastal Regions

地　区 Region	高等学校数（所） Institutions of Higher Education (unit)	本、专科在校学生数（人） Number of Students at School in Undergraduate or Specialized Courses (person)	本、专科毕（结）业生数（人） Number of Students Graduated in Undergraduate or Specialized Courses (person)
全国总计 National Total	**2 442**	**23 913 155**	**6 517 623**
天　津 Tianjin	55	473 114	125 076
河　北 Hebei	113	1 168 796	336 079
辽　宁 Liaoning	112	934 078	247 722
上　海 Shanghai	67	506 596	141 902
江　苏 Jiangsu	153	1 671 173	486 168
浙　江 Zhejiang	102	932 292	252 074
福　建 Fujian	86	701 392	192 804
山　东 Shandong	136	1 658 490	482 579
广　东 Guangdong	137	1 616 838	425 324
广　西 Guangxi	70	629 243	175 760
海　南 Hainan	17	168 270	44 632

10-9 沿海地区卫生基本情况
Basic Conditions of Public Health by Coastal Regions

地　区 Region	卫生机构数（个）Health Institutions (unit)	医疗机构床位数（张）Total Beds (bed)	卫生机构人员（人）Number of Employed Personnel in Health Institutions (person)
全国总计 National Total	**950 297**	**5 724 800**	**9 115 705**
天　津 Tianjin	4 551	53 500	104 201
河　北 Hebei	79 119	284 400	463 283
辽　宁 Liaoning	35 792	231 000	329 679
上　海 Shanghai	4 845	109 800	183 416
江　苏 Jiangsu	31 050	333 100	519 709
浙　江 Zhejiang	30 271	213 300	400 094
福　建 Fujian	27 276	139 300	236 756
山　东 Shandong	68 840	473 800	738 868
广　东 Guangdong	46 534	355 300	662 462
广　西 Guangxi	34 152	168 700	303 759
海　南 Hainan	5 154	30 300	59 285

10-10 沿海地区能源消耗指标（2011年）
Indicators of Energy Consumption by Coastal Regions (2011)

地 区 Region	万元地区生产总值能耗（等价值）Energy Consumption Per 10 000 yuan of GRP (Equivalent Value)		万元工业增加值能耗（规模以上，当量值）Energy Consumption Per 10 000 yuan of Industrial Value-Added Change (above Designated Size, Equivalent Weight)	万元地区生产总值电耗上升或下降 Electricity Consumption per 10 000 yuan of GRP Change（±%）
	指标值（吨标准煤/万元）Index (ton of SCE/10 000 yuan)	上升或下降（±%）Change (±%)		
天 津 Tianjin	0.708	-4.28	-7.48	-7.48
河 北 Hebei	1.300	-3.69	-6.68	-0.36
辽 宁 Liaoning	1.096	-3.40	-5.02	-3.15
上 海 Shanghai	0.618	-5.32	-7.33	-4.42
江 苏 Jiangsu	0.600	-3.52	-5.41	-0.14
浙 江 Zhejiang	0.590	-3.07	-2.40	1.41
福 建 Fujian	0.644	-3.29	-1.16	2.73
山 东 Shandong	0.855	-3.77	-7.67	-0.58
广 东 Guangdong	0.563	-3.78	-5.13	-1.46
广 西 Guangxi	0.800	-3.36	-6.13	-0.28
海 南 Hainan	0.692	5.23	12.53	3.94

注：地区生产总值和工业增加值按2010年价格计算。

Note: Gross regional product and industrial value-added are caculated at 2010 constant prices.

10-11　沿海地区用水情况
Water Use of Coastal Regions

地　区 Region	全年供水总量 （万立方米） Total Annual Volume of Water Supply ($10\ 000\ m^3$)	人均用水量 （立方米/人） Per Capita Water Use (cu.m/person)
全国总计 National Total	**6 462**	
天　津 Tianjin	23.1	167.1
河　北 Hebei	195.3	268.9
辽　宁 Liaoning	142.2	324.3
上　海 Shanghai	116.0	490.6
江　苏 Jiangsu	552.2	698.2
浙　江 Zhejiang	198.1	362.2
福　建 Fujian	200.1	535.8
山　东 Shandong	221.8	229.6
广　东 Guangdong	451.0	427.5
广　西 Guangxi	303.0	649.8
海　南 Hainan	45.3	514.0

10-12 沿海地区全社会固定资产投资

Total Investment in Fixed Assets by the Whole Society of Coastal Regions

单位：亿元 (100 million yuan)

地 区 Region	全社会固定资产投资 Total Investment in Fixed Assets in the Whole Country	#农、林、牧、渔业 Agriculture, Forestry, Animal Husbandry and Fishery	#交通运输、仓储和邮政业 Transport, Storage and Post
全国总计 National Total	**374 694.7**	**10 996.4**	**31 444.9**
天 津 Tianjin	7 934.8	198.7	729.9
河 北 Hebei	19 661.3	804.4	1 543.3
辽 宁 Liaoning	21 836.3	601.4	1 070.1
上 海 Shanghai	5 117.6	11.0	460.8
江 苏 Jiangsu	30 854.2	251.9	1 397.1
浙 江 Zhejiang	17 649.4	200.1	1 349.7
福 建 Fujian	12 439.9	241.9	1 441.9
山 东 Shandong	31 256.0	900.4	1 657.2
广 东 Guangdong	18 751.5	340.6	1 729.7
广 西 Guangxi	9 808.6	370.5	925.8
海 南 Hainan	2 145.4	40.8	143.2

10-13 沿海地区按经营单位所在地分货物进出口总额
Total Value of Imports and Exports by Location of Importers/Exporters of Coastal Regions

单位：万美元 (10 000 USD)

地 区 Region	进出口 Total	出口 Exports	进口 Imports
全国总计 National Total	**386 711 942**	**204 871 442**	**181 840 500**
天 津 Tianjin	11 563 427	4 831 256	6 732 171
河 北 Hebei	5 056 306	2 959 820	2 096 485
辽 宁 Liaoning	10 409 000	5 795 905	4 613 094
上 海 Shanghai	43 658 695	20 673 017	22 985 679
江 苏 Jiangsu	54 796 149	32 852 352	21 943 797
浙 江 Zhejiang	31 240 136	22 451 714	8 788 421
福 建 Fujian	15 593 796	9 783 259	5 810 536
山 东 Shandong	24 554 432	12 870 921	11 683 512
广 东 Guangdong	98 402 046	57 405 077	40 996 969
广 西 Guangxi	2 948 446	1 546 775	1 401 671
海 南 Hainan	1 432 210	313 610	1 118 600

10-14 沿海地区城镇居民平均每人全年家庭收入和消费性支出
Per Capita Annual Income and Consumption Expenditure of Urban Households by Coastal Regions

单位：元 (yuan)

地　区 Region	可支配收入 Disposable Income	总收入 Total Income	现金消费支出 Cash Consumption Expenditure
全国平均水平 National Average	**24 564.72**	**26 958.99**	**16 674.32**
天　津 Tianjin	29 626.41	32 944.01	20 024.24
河　北 Hebei	20 543.44	21 899.42	12 531.12
辽　宁 Liaoning	23 222.67	25 915.72	16 593.60
上　海 Shanghai	40 188.34	44 754.50	26 253.47
江　苏 Jiangsu	29 676.97	32 519.10	18 825.28
浙　江 Zhejiang	34 550.30	37 994.83	21 545.18
福　建 Fujian	28 055.24	30 877.92	18 593.21
山　东 Shandong	25 755.19	28 005.61	15 778.24
广　东 Guangdong	30 226.71	34 044.38	22 396.35
广　西 Guangxi	21 242.80	23 209.41	14 243.98
海　南 Hainan	20 917.71	22 809.87	14 456.55

10-15 沿海地区农村居民家庭人均纯收入和生活消费支出
Per Capita Annual Net Income and Consumption Expenditure of Rural Households by Coastal Regions

单位：元　　　　　　　　　　　　　　　　　　　　　　　　　　　　　　　　　　(yuan)

地　区 Region	纯收入 Net Income	生活消费支出 Consumption Expenditure
全国平均水平 **National Average**	**7 916.58**	**5 908.02**
天　津 Tianjin	14 025.54	8 336.55
河　北 Hebei	8 081.39	5 364.14
辽　宁 Liaoning	9 383.72	5 998.39
上　海 Shanghai	17 803.68	11 971.50
江　苏 Jiangsu	12 201.95	9 138.18
浙　江 Zhejiang	14 551.92	10 652.73
福　建 Fujian	9 967.17	7 401.92
山　东 Shandong	9 446.54	6 775.95
广　东 Guangdong	10 542.84	7 458.56
广　西 Guangxi	6 007.55	4 933.58
海　南 Hainan	7 408.00	4 776.30

10-16 沿海地区年末人口数
Year-end Population by Coastal Regions

单位：万人 (10 000 persons)

地 区 Region	2006	2007	2008	2009	2010	2011	2012
全国总计 National Total	**131 448**	**132 129**	**132 802**	**133 450**	**134 091**	**134 735**	**135 404**
天 津 Tianjin	1 705	1 115	1 176	1 228	1 299	1 355	1 413
河 北 Hebei	6 898	6 943	6 989	7 034	7 194	7 241	7 288
辽 宁 Liaoning	4 271	4 298	4 315	4 341	4 375	4 383	4 389
上 海 Shanghai	1 964	2 064	2 141	2 210	2 303	2 347	2 380
江 苏 Jiangsu	7 656	7 723	7 762	7 810	7 869	7 899	7 920
浙 江 Zhejiang	5 072	5 155	5 212	5 276	5 447	5 463	5 477
福 建 Fujian	3 585	3 612	3 639	3 666	3 693	3 720	3 748
山 东 Shandong	9 309	9 367	9 417	9 470	9 588	9 637	9 685
广 东 Guangdong	9 442	9 660	9 893	10 130	10 441	10 505	10 594
广 西 Guangxi	4 719	4 768	4 816	4 856	4 610	4 645	4 682
海 南 Hainan	836	845	854	864	869	877	887

注:2010年数据为当年人口普查数据推算数；其余年份数据为年度人口抽样调查推算数据。各地区数据为常住人口口径。

Note: Data of 2010 are the census year estimates; the rest are the estimates from the annual national sample survey on population changes. Data by region are usual residents.

10-17 沿海城市人口情况（2011年）
Population of Coastal Cities, 2011

单位：万人 (10 000 persons)

沿海城市	Coastal City	常住人口 Permanent Population	年底总人口 Total Population by the End of the Year
合 计	**Total**	**28 558.1**	**24 456.7**
天 津	**Tianjin**	**1 354.6**	**996.4**
河 北	**Hebei**	**1 783.1**	**1 761.7**
唐 山	Tangshan	762.7	737.1
秦皇岛	Qinhuangdao	300.6	289.8
沧 州	Cangzhou	719.8	734.8
辽 宁	**Liaoning**	**1 785.9**	**1 785.9**
大 连	Dalian	588.5	588.5
丹 东	Dandong	241.1	241.1
锦 州	Jinzhou	308.3	308.3
营 口	Yingkou	235.5	235.5
盘 锦	Panjin	131.2	131.2
葫芦岛	Huludao	281.3	281.3
上 海	**Shanghai**	**2 347.5**	**1 419.4**
江 苏	**Jiangsu**	**1 891.2**	**2 090.8**
南 通	Nantong	728.9	764.9
连云港	Lianyungang	438.6	505.2
盐 城	Yancheng	723.7	820.7

10-17 续表1 continued

沿海城市	Coastal City	常住人口 Permanent Population	年底总人口 Total Population by the End of the Year
浙　江	**Zhejiang**	**4 211.0**	**3 537.4**
杭　州	Hangzhou	873.8	695.7
宁　波	Ningbo	762.8	576.4
温　州	Wenzhou	914.3	798.4
嘉　兴	Jiaxing	453.1	343.1
绍　兴	Shaoxing	493.4	440.0
舟　山	Zhoushan	113.7	97.0
台　州	Taizhou	599.9	586.8
福　建	**Fujian**	**2 948.0**	**2 669.9**
福　州	Fuzhou	720.0	649.4
厦　门	Xiamen	361.0	185.3
莆　田	Putian	279.0	326.5
泉　州	Quanzhou	821.0	689.5
漳　州	Zhangzhou	484.0	479.2
宁　德	Ningde	283.0	340.0
山　东	**Shandong**	**3 637.2**	**3 405.3**
青　岛	Qingdao	879.5	766.4
东　营	Dongying	205.5	186.0
烟　台	Yantai	697.6	651.8
潍　坊	Weifang	915.5	877.6
威　海	Weihai	280.1	253.8
日　照	Rizhao	281.9	289.0
滨　州	Binzhou	377.1	380.7

10-17 续表2 continued

沿海城市	Coastal City	常住人口 Permanent Population	年底总人口 Total Population by the End of the Year
广 东	**Guangdong**	**7 765.1**	**5 918.9**
广 州	Guangzhou	1 275.1	814.6
深 圳	Shenzhen	1 046.7	279.4
珠 海	Zhuhai	156.8	106.0
汕 头	Shantou	541.7	529.4
江 门	Jiangmen	446.6	393.7
湛 江	Zhanjiang	706.9	792.1
茂 名	Maoming	588.3	761.3
惠 州	Huizhou	463.4	343.0
汕 尾	Shanwei	295.5	347.2
阳 江	Yangjiang	244.5	284.6
东 莞	Dongguan	825.5	184.8
中 山	Zhongshan	314.2	150.7
潮 州	Chaozhou	268.4	262.8
揭 阳	Jieyang	591.5	669.3
广 西	**Guangxi**	**554.2**	**650.5**
北 海	Beihai	155.4	167.9
防城港	Fangchenggang	87.8	91.4
钦 州	Qinzhou	311.0	391.2
海 南	**Hainan**	**280.3**	**220.5**
海 口	Haikou	209.7	162.4
三 亚	Sanya	70.6	58.1

注：本表各省数据为合计数。

Note: The data for the provinces are the totals.

10-18 沿海县人口情况（2011年）
Population of Coastal Counties, 2011

单位：万人 (10 000 persons)

沿海县 Coastal County		年末总人口 Total Population by the End of the Year	乡村人口 Rural Population
合　计	**Total**	**8 663.90**	**7 099.20**
河　北	**Hebei**	**296.50**	**257.20**
滦　南	Luannan	58.20	54.00
乐　亭	Leting	49.40	43.50
唐　海	Tanghai	14.40	11.20
昌　黎	Changli	55.90	49.10
抚　宁	Funing	49.40	43.00
黄　骅	Huanghua	45.80	37.80
海　兴	Haixing	23.40	18.60
辽　宁	**Liaoning**	**662.90**	**514.60**
长　海	Changhai	7.30	6.30
瓦房店	Wafangdian	100.30	73.00
普兰店	Pulandian	93.20	58.10
庄　河	Zhuanghe	90.70	66.40
东　港	Donggang	61.00	50.10
凌　海	Linghai	52.90	43.90
盖　州	Gaizhou	72.30	62.00
大　洼	Dawa	38.30	30.70
盘　山	Panshan	27.80	27.60
绥　中	Suizhong	64.00	55.00
兴　城	Xingcheng	55.10	41.50

10-18 续表1 continued

沿海县 Coastal County		年末总人口 Total Population by the End of the Year	乡村人口 Rural Population
江　苏	**Jiangsu**	**1 285.50**	**1 020.00**
海　安	Hai'an	93.70	72.00
如　东	Rudong	104.80	87.00
启　东	Qidong	112.40	91.00
海　门	Haimen	100.00	80.00
赣　榆	Ganyu	114.10	88.00
东　海	Donghai	116.20	96.00
灌　云	Guanyun	101.80	81.00
灌　南	Guannan	77.50	63.00
响　水	Xiangshui	61.60	47.00
滨　海	Binhai	120.10	94.00
射　阳	Sheyang	97.40	76.00
东　台	Dongtai	113.40	91.00
大　丰	Dafeng	72.50	54.00
浙　江	**Zhejiang**	**1 481.50**	**1 308.30**
象　山	Xiangshan	54.20	39.10
宁　海	Ninghai	61.50	51.60
余　姚	Yuyao	83.50	73.30
慈　溪	Cixi	104.20	122.70
奉　化	Fenghua	48.40	39.10
洞　头	Dongtou	12.90	8.30
平　阳	Pingyang	87.50	66.50
苍　南	Cangnan	131.70	98.50
瑞　安	Rui'an	121.10	113.00
乐　清	Yueqing	126.00	120.90
海　盐	Haiyan	37.50	34.60
海　宁	Haining	66.30	52.90
平　湖	Pinghu	48.80	34.60

10-18 续表2 continued

沿海县 Coastal County		年末总人口 Total Population by the End of the Year	乡村人口 Rural Population
绍　兴	Shaoxing	72.60	82.10
上　虞	Shangyu	77.70	64.30
岱　山	Daishan	19.10	14.70
嵊　泗	Shengsi	7.90	4.60
玉　环	Yuhuan	42.00	60.20
三　门	Sanmen	42.90	34.20
温　岭	Wenling	119.30	104.50
临　海	Linhai	116.40	88.60
福　建	**Fujian**	**1 292.60**	**1 101.10**
连　江	Lianjiang	63.60	56.80
罗　源	Luoyuan	25.60	22.60
平　潭	Pingtan	40.70	36.10
福　清	Fuqing	128.20	116.00
长　乐	Changle	69.00	62.60
仙　游	Xianyou	109.10	98.70
惠　安	Hui'an	96.50	67.70
石　狮	Shishi	31.80	24.70
晋　江	Jinjiang	107.00	111.20
南　安	Nan'an	151.00	112.20
云　霄	Yunxiao	43.80	35.20
漳　浦	Zhangpu	85.90	79.40
诏　安	Zhao'an	60.70	55.70
东　山	Dongshan	21.00	14.00
龙　海	Longhai	82.40	70.60
霞　浦	Xiapu	53.30	43.30
福　安	Fu'an	64.80	46.70
福　鼎	Fuding	58.20	47.60

10-18 续表3 continued

沿海县 Coastal County		年末总人口 Total Population by the End of the Year	乡村人口 Rural Population
山　东	**Shandong**	**1 226.60**	**1 002.90**
胶　州	Jiaozhou	80.70	63.10
即　墨	Jimo	113.20	96.30
胶　南	Jiaonan	84.10	70.10
垦　利	Kenli	23.00	18.00
利　津	Lijin	30.00	26.00
广　饶	Guangrao	50.00	43.00
长　岛	Changdao	4.30	2.60
龙　口	Longkou	63.50	48.90
莱　阳	Laiyang	87.40	75.10
莱　州	Laizhou	85.80	69.20
蓬　莱	Penglai	45.00	37.30
招　远	Zhaoyuan	57.10	45.30
海　阳	Haiyang	66.50	58.80
寿　光	Shouguang	104.60	88.40
昌　邑	Changyi	58.20	49.90
文　登	Wendeng	64.00	47.00
荣　成	Rongcheng	67.10	44.50
乳　山	Rushan	57.10	47.40
无　棣	Wudi	46.00	37.00
沾　化	Zhanhua	39.00	35.00
广　东	**Guangdong**	**1 752.60**	**1 423.60**
南　澳	Nan'ao	7.40	5.60
台　山	Taishan	98.70	84.60
恩　平	Enping	50.00	33.00
遂　溪	Suixi	106.80	93.20
徐　闻	Xuwen	73.60	59.00

10-18 续表4 continued

沿海县 Coastal County		年末总人口 Total Population by the End of the Year	乡村人口 Rural Population
廉　江	Lianjiang	172.20	135.40
雷　州	Leizhou	171.20	152.60
吴　川	Wuchuan	112.20	86.40
电　白	Dianbai	146.00	119.20
惠　东	Huidong	85.20	55.50
海　丰	Haifeng	82.00	68.00
陆　丰	Lufeng	180.00	112.00
阳　西	Yangxi	52.20	42.30
阳　东	Yangdong	49.10	46.90
饶　平	Raoping	102.00	89.00
揭　东	Jiedong	130.00	119.60
惠　来	Huilai	134.00	121.30
广　西	**Guangxi**	**118.70**	**72.50**
东　兴	Dongxing	13.30	9.30
合　浦	Hepu	105.40	63.20
海　南	**Hainan**	**547.00**	**399.00**
琼　海	Qionghai	50.00	36.00
儋　州	Danzhou	104.00	64.00
文　昌	Wenchang	59.00	46.00
万　宁	Wanning	61.00	44.00
东　方	Dongfang	47.00	34.00
澄　迈	Chengmai	57.00	42.00
临　高	Lingao	50.00	43.00
昌　江	Changjiang	27.00	17.00
乐　东	Ledong	54.00	44.00
陵　水	Lingshui	38.00	29.00

注：本表各省数据为合计数。

Note: The data for the provinces are the totals.

10-19 沿海地区城镇单位就业人员情况
Employed Persons in Urban Units by Coastal Regions

单位：万人 (10 000 persons)

地 区 Region	2011	2012
全国总计 National Total	**14 413.3**	**15 236.4**
天 津 Tianjin	268.2	289.1
河 北 Hebei	555.4	619.9
辽 宁 Liaoning	579.6	598.7
上 海 Shanghai	497.3	555.7
江 苏 Jiangsu	811.3	830.9
浙 江 Zhejiang	995.7	1 070.1
福 建 Fujian	596.3	637.9
山 东 Shandong	1 050.4	1 110.2
广 东 Guangdong	1 238.2	1 304.0
广 西 Guangxi	341.6	358.0
海 南 Hainan	85.1	90.1

10-20 沿海城市就业人员情况（2011年）
Number of Employed Persons by Coastal Cities, 2011

单位：万人 (10 000 persons)

沿海城市 Coastal City	就业人员 Number of Employed Persons	城镇就业人员 Urban Employed Persons
合 计 Total	**17 071.16**	**4 281.61**
天 津 Tianjin	**763.16**	**268.24**
河 北 Hebei	**1 011.33**	**165.76**
唐 山 Tangshan	440.17	87.30
秦皇岛 Qinhuangdao	161.57	29.88
沧 州 Cangzhou	409.59	48.58
辽 宁 Liaoning	**1 112.98**	**259.75**
大 连 Dalian	437.20	109.81
丹 东 Dandong	125.77	26.80
锦 州 Jinzhou	160.32	26.85
营 口 Yingkou	152.77	26.63
盘 锦 Panjin	100.41	47.44
葫芦岛 Huludao	136.51	22.22
上 海 Shanghai	**1 104.33**	**497.32**
江 苏 Jiangsu	**1 130.54**	**153.93**
南 通 Nantong	464.88	65.99
连云港 Lianyungang	308.18	35.16
盐 城 Yancheng	357.48	52.78
浙 江 Zhejiang	**2 811.96**	**854.05**
杭 州 Hangzhou	637.77	264.30
宁 波 Ningbo	493.83	171.80
温 州 Wenzhou	575.89	111.31
嘉 兴 Jiaxing	321.39	79.05
绍 兴 Shaoxing	343.28	125.18
舟 山 Zhoushan	68.99	16.10
台 州 Taizhou	370.81	86.31
福 建 Fujian	**1 905.03**	**509.93**
福 州 Fuzhou	425.59	128.07
厦 门 Xiamen	251.24	110.50
莆 田 Putian	194.54	34.02

10-20 续表 continued

沿海城市 Coastal City		就业人员 Number of Employed Persons	城镇就业人员 Urban Employed Persons
泉　州	Quanzhou	570.67	173.96
漳　州	Zhangzhou	289.21	45.70
宁　德	Ningde	173.78	17.68
山　东	**Shandong**	**2 227.40**	**468.80**
青　岛	Qingdao	547.60	128.40
东　营	Dongying	126.00	46.10
烟　台	Yantai	434.10	106.00
潍　坊	Weifang	513.50	68.70
威　海	Weihai	173.30	56.60
日　照	Rizhao	182.50	21.40
滨　州	Binzhou	250.40	41.60
广　东	**Guangdong**	**4 482.59**	**1 021.40**
广　州	Guangzhou	743.10	310.24
深　圳	Shenzhen	764.54	263.71
珠　海	Zhuhai	104.09	67.00
汕　头	Shantou	238.55	35.59
江　门	Jiangmen	253.03	54.87
湛　江	Zhanjiang	329.13	42.86
茂　名	Maoming	275.58	35.20
惠　州	Huizhou	267.94	84.75
汕　尾	Shanwei	119.23	15.76
阳　江	Yangjiang	137.87	18.87
东　莞	Dongguan	628.54	24.87
中　山	Zhongshan	208.64	31.27
潮　州	Chaozhou	138.80	13.36
揭　阳	Jieyang	273.55	23.05
广　西	**Guangxi**	**372.02**	**39.27**
北　海	Beihai	80.70	14.30
防城港	Fangchenggang	57.33	9.75
钦　州	Qinzhou	233.99	15.22
海　南	**Hainan**	**149.82**	**43.16**
海　口	Haikou	118.15	35.52
三　亚	Sanya	31.67	7.64

注：本表各省数据为合计数。

Note: The data for the provinces are the totals.

10-21 沿海县就业人员情况（2011年）
Number of Employed Persons of Coastal Counties, 2011

单位：人 (person)

沿海县 Coastal County		年末单位从业人员数 Employed Persons by the End of the year	乡村从业人员数 Rural Employees
合　计	**Total**	**8 899 380**	**38 980 299**
河　北	**Hebei**	**203 133**	**1 427 220**
丰　南	Fengnan		
滦　南	Luannan	30 847	292 727
乐　亭	Leting	25 387	264 829
唐　海	Tanghai	43 700	69 526
昌　黎	Changli	22 671	284 752
抚　宁	Funing	35 579	230 328
黄　骅	Huanghua	32 888	180 762
海　兴	Haixing	12 061	104 296
辽　宁	**Liaoning**	**572 483**	**2 801 251**
长　海	Changhai	9 574	30 658
瓦房店	Wafangdian	60 784	365 196
普兰店	Pulandian	64 800	357 166
庄　河	Zhuanghe	44 475	372 998
东　港	Donggang	43 372	268 367
凌　海	Linghai	30 994	241 801
盖　州	Gaizhou	22 330	333 656
大　洼	Dawa	182 549	187 869
盘　山	Panshan	64 160	159 358
绥　中	Suizhong	21 257	281 039
兴　城	Xingcheng	28 188	203 143

10-21 续表1 continued

沿海县 Coastal County		年末单位从业人员数 Employed Persons by the End of the year	乡村从业人员数 Rural Employees
江　苏	**Jiangsu**	**686 299**	**5 298 200**
海　安	Hai'an	72 117	389 200
如　东	Rudong	70 443	484 700
启　东	Qidong	67 525	528 800
海　门	Haimen	66 597	499 200
赣　榆	Ganyu	39 255	422 500
东　海	Donghai	40 997	465 500
灌　云	Guanyun	32 324	373 100
灌　南	Guannan	35 309	318 300
响　水	Xiangshui	33 191	218 800
滨　海	Binhai	39 552	441 600
射　阳	Sheyang	56 220	356 200
东　台	Dongtai	69 947	484 000
大　丰	Dafeng	62 822	316 300
浙　江	**Zhejiang**	**2 490 252**	**8 143 000**
象　山	Xiangshan	298 841	275 300
宁　海	Ninghai	79 139	332 300
余　姚	Yuyao	119 114	455 800
慈　溪	Cixi	159 945	800 000
奉　化	Fenghua	61 525	270 700
洞　头	Dongtou	7 958	53 600
平　阳	Pingyang	77 281	406 300
苍　南	Cangnan	95 008	619 100
瑞　安	Rui'an	115 310	650 300
乐　清	Yueqing	236 261	693 800
海　盐	Haiyan	68 349	216 700
海　宁	Haining	127 034	312 000
平　湖	Pinghu	140 727	223 300

10-21 续表2 continued

沿海县 Coastal County		年末单位从业人员数 Employed Persons by the End of the year	乡村从业人员数 Rural Employees
绍　兴	Shaoxing	287 163	497 900
上　虞	Shangyu	221 964	377 700
岱　山	Daishan	18 400	87 500
嵊　泗	Shengsi	11 900	27 000
玉　环	Yuhuan	106 655	396 100
三　门	Sanmen	31 698	218 600
温　岭	Wenling	95 429	652 500
临　海	Linhai	130 551	576 500
福　建	**Fujian**	**1 774 784**	**5 986 667**
连　江	Lianjiang	43 283	295 562
罗　源	Luoyuan	16 816	99 735
平　潭	Pingtan	20 569	186 288
福　清	Fuqing	204 585	554 740
长　乐	Changle	50 585	278 896
仙　游	Xianyou	38 056	506 990
惠　安	Hui'an	228 596	369 610
石　狮	Shishi	159 312	137 757
晋　江	Jinjiang	613 018	729 827
南　安	Nan'an	106 915	750 206
云　霄	Yunxiao	24 396	173 621
漳　浦	Zhangpu	61 615	447 890
诏　安	Zhao'an	29 935	326 708
东　山	Dongshan	18 950	80 212
龙　海	Longhai	89 472	397 579
霞　浦	Xiapu	17 844	219 272
福　安	Fu'an	28 331	173 264
福　鼎	Fuding	22 506	258 510

10-21 续表3 continued

沿海县 Coastal County		年末单位从业人员数 Employed Persons by the End of the year	乡村从业人员数 Rural Employees
山　东	**Shandong**	**1 645 120**	**5 600 061**
胶　州	Jiaozhou	165 681	363 165
即　墨	Jimo	156 892	561 591
胶　南	Jiaonan	194 590	366 312
垦　利	Kenli	31 345	86 578
利　津	Lijin	17 256	147 092
广　饶	Guangrao	74 231	274 834
长　岛	Changdao	5 306	12 704
龙　口	Longkou	94 582	275 228
莱　阳	Laiyang	83 263	422 886
莱　州	Laizhou	107 295	385 751
蓬　莱	Penglai	87 740	200 596
招　远	Zhaoyuan	84 489	224 260
海　阳	Haiyang	38 502	357 513
寿　光	Shouguang	84 545	412 470
昌　邑	Changyi	39 297	267 848
文　登	Wendeng	107 437	279 480
荣　成	Rongcheng	151 099	233 337
乳　山	Rushan	53 519	266 627
无　棣	Wudi	37 204	243 749
沾　化	Zhanhua	30 847	218 040
广　东	**Guangdong**	**1 004 423**	**7 289 965**
南　澳	Nan'ao	4 919	24 021
台　山	Taishan	50 583	494 121
恩　平	Enping	63 576	187 140
遂　溪	Suixi	40 875	448 321
徐　闻	Xuwen	40 356	316 107

10-21 续表4 continued

沿海县 Coastal County		年末单位从业人员数 Employed Persons by the End of the year	乡村从业人员数 Rural Employees
廉　江	Lianjiang	60 474	697 022
雷　州	Leizhou	54 439	688 642
吴　川	Wuchuan	49 357	535 839
电　白	Dianbai	58 600	564 313
惠　东	Huidong	61 182	373 688
海　丰	Haifeng	40 587	396 620
陆　丰	Lufeng	48 407	623 864
阳　西	Yangxi	283 579	238 177
阳　东	Yangdong	29 593	262 809
饶　平	Raoping	30 171	435 063
揭　东	Jiedong	47 908	601 852
惠　来	Huilai	39 817	402 366
广　西	**Guangxi**	**53 711**	**401 700**
东　兴	Dongxing	8 186	54 800
合　浦	Hepu	45 525	346 900
海　南	**Hainan**	**469 175**	**2 032 235**
琼　海	Qionghai	45 680	191 635
儋　州	Danzhou	36 030	320 346
文　昌	Wenchang	65 232	233 782
万　宁	Wanning	89 041	207 013
东　方	Dongfang	25 137	172 563
澄　迈	Chengmai	122 094	240 068
临　高	Lingao	15 989	198 125
昌　江	Changjiang	20 124	89 457
乐　东	Ledong	35 688	240 977
陵　水	Lingshui	14 160	138 269

注：本表各省数据为合计数，缺少福建金门县数据。

Note: The data for the provinces are the totals, and lacking the data for Jinmen of Fujian Province.

主要统计指标解释

1. 国内(或地区)生产总值 指一个国家（或地区）所有常驻单位在一定时期内生产活动的最终成果。国内生产总值有三种表现形态，即价值形态、收入形态和产品形态。从价值形态看，它是所有常驻单位在一定时期内生产的全部货物和服务价值超过同期中间投入的全部非固定资产货物和服务价值的差额，即所有常驻单位的增加值之和；从收入形态看，它是所有常驻单位在一定时期内创造并分配给常驻单位和非常驻单位的初次收入分配之和；从产品形态看，它是所有常驻单位在一定时期内最终使用的货物和服务价值与货物和服务净出口价值之和。在实际核算中，国内生产总值有三种计算方法，即生产法、收入法和支出法。三种方法分别从不同的方面反映国内生产总值及其构成。

2. 三次产业 是根据社会生产活动历史发展的顺序对产业结构的划分，产品直接取自自然界的部门称为第一产业，对初级产品进行再加工的部门称为第二产业，为生产和消费提供各种服务的部门称为第三产业。它是世界上较为通用的产业结构分类，但各国的划分不尽一致。我国的三次产业划分是：

第一产业：是指农、林、牧、渔业。

第二产业：是指采矿业，制造业，电力、燃气及水的生产和供应业，建筑业。

第三产业：是指除第一、二产业以外的其他行业。第三产业包括：交通运输、仓储和邮政业，信息传输、计算机服务和软件业，批发和零售业，住宿和餐饮业，金融业，房地产业，租赁和商务服务业，科学研究、技术服务和地质勘查业，水利、环境和公共设施管理业，居民服务和其他服务业，教育，卫生、社会保障和社会福利业，文化、体育和娱乐业，公共管理和社会组织，国际组织。

3. 增加值 是指各行各业生产经营和劳务活动的最终成果，采用生产法和收入法两种方法计算。

生产法 是从货物和服务活动在生产过程中形成的总产品入手，剔除生产过程中投入的中间产品价值，得到新增价值的方法。

收入法 又称分配法。按收入法计算国内生产总值是从生产过程创造的收入的角度对常驻单位的生产活动成果进行核算；按照此法计算，增加值由劳动者报酬、固定资产折旧、生产税净额和营业盈余四个部分组成。

4. 年末总人口 是指每年 12 月 31 日 24 时一定地区范围内的有生命的个人的人口总和。

5. 财政收入 指国家财政参与社会产品分配所取得的收入，是实现国家职能的财力保证。财政收入所包括的内容几经变化，目前主要包括：

(1)各项税收：包括增值税、营业税、消费税、土地增值税、城市维护建设税、资源税、城市土地使用税、企业所得税、个人所得税、关税、证券交易印花税、车辆购置税、农牧业税和耕地占用税等。

(2)专项收入：包括排污费收入、城市水资源费收入、矿产资源补偿费收入、教育费附加收入等。

(3)其他收入：包括利息收入、基本建设贷款归还收入、基本建设收入、捐赠收入等。

(4)国有企业亏损补贴：此项为负收入，冲减财政收入。主要包括对工业企业、商业企业、粮食企业的补贴。

6. 财政支出 国家财政将筹集起来的资金进行分配使用，以满足经济建设和各项事业的需要。

7. 基本建设支出 指按国家有关规定，属于基本建设范围内的基本建设有偿使用、拨款、资本金支出以及经国家批准对专项和政策性基建投资贷款，在部门的基建投资额中统筹支付的贴息支出。

8. 普通高等学校 指按照国家规定的设置标准和审批程序批准举办的，通过全国普通高等学校统一招生考试，招收高中毕业生为主要培养对象，实施高等教育的全日制大学、独立设置的学院和高等专科学校、高等职业学校和其他机构。

大学、独立设置的学院主要实施本科层次以上教育，高等专科学校、高等职业学校实施专科层次教育，其他机构是承担国家普通招生计划任务不计校数的机构。包括普通高等学校分校和批准筹建的普通高等学校等。

9. 卫生机构 包括医疗机构、疾病预防控制中心(防疫站)、采供血机构、卫生监督及监测(检验)机构、医学科研和在职培训机构、健康教育所等。

10. 医疗机构 包括医院、社区卫生服务中心(站)、疗养院、卫生院、门诊部、诊所(卫生所、医务室)、妇幼保健院(所、站)、专科疾病防治院(所、站)、急救中心(站)和临床检验中心。医疗机构分为非赢利性医疗机构和赢利性医疗机构。

11. 单位国内生产总值能耗 指一定时期内，一个国家或地区每生产一个单位的国内生产总值所消耗的能源。计算公式为：

$$单位国内生产总值能源=\frac{能源消费总量}{国内生产总值}$$

12. 单位国内生产总值电耗 指一定时期内，一个国家或地区每生产一个单位的国内生产总值所消耗的电力。计算公式为：

$$单位国内生产总值电耗=\frac{全社会用电量}{国内生产总值}$$

13. 单位工业增加值能耗 指一定时期内，一个国家或地区每生产一个单位的工业增加值所消耗的能源。计算公式为：

$$单位工业增加值能耗=\frac{工业能源消费总量}{工业增加值}$$

14. 用水总量 指分配给各类用户的包括输水损失在内的毛用水量之和，不包括海水直接利用量。

15. 全社会固定资产投资 是以货币形式表现的在一定时期内全社会建造和购置固定资产的工作量以及与此有关的费用的总称。该指标是反映固定资产投资规模、结构和发展速度的综合性指标,又是观察工程进度和考核投资效果的重要依据。全社会固定资产投资按登记注册类型可分为国有、集体、个体、联营、股份制、外商、港澳台商、其他等。

16. 进出口总额 指实际进出我国国境的货物总金额。包括对外贸易实际进出口货物，来料加工装配进出口货物，国家间、联合国及国际组织无偿援助物资和赠送品，华侨、港澳台同胞和外籍华人捐赠品，租赁期满归承租人所有的租赁货物，进料加工进出口货物，边境地方贸易及边境地区小额贸易进出口货物(边民互市贸易除外)，中外合资企业、中外合作经营企业、外商独资经营企业进出口货物和公用物品，到、离岸价格在规定限额以上的进出口货样和广告品(无商业价值、无使用价值和免费提供出口的除外)，从保税仓库提取在中国境内销售的进口货物，以及其他进出口货物。该指标可以观察一个国家在对外贸易方面的总规模。我国规定出口货物按离岸价格统计，进口货物按到岸价格统计。

17. 商品经营单位所在地进、出口额　指在所在地海关注册登记的有进出口经营权的企业实际进、出口额。

18. 城镇家庭可支配收入　指家庭成员得到可用于最终消费支出和其它非义务性支出以及储蓄的总和，即居民家庭可以用来自由支配的收入。它是家庭总收入扣除交纳的所得税、个人交纳的社会保障支出以及记账补贴后的收入。计算公式为：

可支配收入=家庭总收入-交纳所得税-个人交纳的社会保障支出-记账补贴

19. 城镇家庭消费性支出　指家庭用于日常生活的支出，包括食品、衣着、家庭设备用品及服务、医疗保健、交通和通信、娱乐教育文化服务、居住、杂项商品和服务等八大类支出。

20. 总收入　指调查期内农村住户和住户成员从各种来源渠道得到的收入总和。按收入的性质划分为工资性收入、家庭经营收入、财产性收入和转移性收入。

21. 纯收入　指农村住户当年从各个来源得到的总收入相应地扣除所发生的费用后的收入总和。计算方法：

纯收入=总收入-税费支出-家庭经营费用支出-生产性固定资产折旧-赠送农村亲友支出

纯收入主要用于再生产投入和当年生活消费支出，也可用于储蓄和各种非义务性支出。“农民人均纯收入”按人口平均的纯收入水平，反映的是一个地区或一个农户农村居民的平均收入水平。

22. 人口数　指一定时点、一定地区范围内有生命的个人总和。

年度统计的年末人口数指每年 12 月 31 日 24 时的人口数。年度统计的全国人口总数内未包括香港、澳门特别行政区和台湾省以及海外华侨人数。

23. 城镇人口　城镇人口是指居住在城镇范围内的全部常住人口

24. 就业人员　指在 16 周岁及以上，从事一定社会劳动并取得劳动报酬或经营收入的人员。这一指标反映了一定时期内全部劳动力资源的实际利用情况，是研究我国基本国情国力的重要指标。

Explanatory Notes on Main Statistical Indicators

1. Gross Domestic Product (GDP) refers to the final result of the primary distribution of the income created by all the resident units of a country (or a region) during a certain period of time. Gross domestic product is expressed in three different forms, i.e. value, income, and products respectively. The form of value refers to the total value of all products and services produced by all resident units during a certain period of time minus total value of intermediate input of materials and services of the nature of non-fixed assets or the summation of the value added of all resident units: the form of income includes all the income created by all resident units and distributed primarily to all resident and non-resident units; the form of products refers to the summation of the value of the products and services finally used and the net export value of products and services by all resident units during a given period of time. In the practice of national accounting, gross domestic product is calculated with three approaches, i.e. production approach, income approach, and expenditure approach, which reflect the gross domestic product and its composition from different aspects.

2. Three Industries Industrial structure is classified according to the sequence of historical development of social productive activities. Primary industry refers to the extraction of natural resources; secondary industry involves processing of primary products; and tertiary industry provides services of various kinds for production and consumption. The above classification is universal in the world although it varies to some extent from country to country. The three industries in China are divided as follows:

Primary industry: refers to farming, forestry, sideline production and fishery.

Secondary industry: refers to such industries as mining, manufacturing, production and supply of electric power, fuel gas and water, and construction.

Tertiary industry: refers to all the other industries not included in the primary or secondary industries. It includes such industries as communications and transportation, storage and postal service; information transmission, computer service and software; wholesale and retailing; accommodation and catering; financial service; real estate; charter business and commercial affairs service; scientific research, technological service and geological survey; water conservancy, environmental and other public facilities management; residents service and other service trades; education, health, social security and social welfare; culture, sports and entertainment business as well as public administration and social organizations, and international organizations.

3. Added Value refers to the final result of production operation and labor activities of all trades and professions, which is calculated by using the methods of production and income.

Production Method: refers to the method whereby to get the newly added value by proceeding from the gross product of goods and service activities occurring in the course of production and then rejecting the value of intermediate product input in the course of production.

Income Method: is also called the distribution method. The calculation of the gross domestic product (GDP) by the income method is the accounting of the result of productive activities of permanent units from the angle of the income created in the course of production. According to this method, the added value is composed of the payment for laborers, depreciation for fixed assets, net tax on production and business surplus.

4. Total Population by the End of the Year refers to the sum of living individuals within a particular range of area at 24:00 on December 31 of each year.

5. Government Revenue refers to the income obtained by the government finance through participating in the distribution of social products. It is the financial guarantee to ensure government functioning. The contents of government revenue have changed several times. Now it includes the following main items:

(1) Various tax revenues, including value added tax, business tax, consumption tax, land

value-added tax, tax on city maintenance and construction, resources tax, tax on use of urban land, enterprise income tax, personal income tax, tariff, stamp tax on security transactions, tax on purchase of motor vehicles, tax on agriculture and animal husbandry and tax on occupancy of cultivated land, etc.

(2) Special revenues, including revenues from the fee on sewage treatment, fee on urban water resources, fee for the compensation of mineral resources and extra-charges for education, etc.

(3) Other revenues, including revenues from interest, repayment of capital construction loan, capital construction projects, and donations and grants.

(4) Subsidies for the losses of State-owned enterprises. This is an item of negative revenue, counteracting revenues and consisting of subsidies to industrial, commercial and grain purchasing and supply enterprises.

6. Government Expenditure refers to the distribution and use of the funds which the government finance has raised, so as to meet the needs of economic construction and various causes.

7. Expenditure for capital construction It refers to the non-gratuitous use of, appropriation of funds for and capital outlay on capital construction in the area of capital construction. It also covers the loans on capital construction approved by the government for special purposes or policy purposes and the expenditure with discount paid in an overall way within the amount of the funds appropriated to the departments for capital construction.

8. Regular Institutions of Higher Learning refer to educational establishments set up according to the government evaluation and approval procedures, enrolling graduates from senior secondary schools and providing higher education courses and training for senior professionals. They include full-time universities, colleges, institutions of higher professional education, institutions of higher vocational education and others.

Universities and colleges primarily provide undergraduate courses; institutions of higher professional education and institutions of higher vocational education primarily provide professional trainings; and others refer to educational establishments, which are responsible for enrolling higher education students under the State Plan but not enumerated in the total number of schools, including: branch schools of universities and colleges, and universities and colleges that have been approved and under plan for construction.

9. Health Care Institutions include: medical institutions, disease prevention and control centres (epidemic prevention stations), blood gathering and supplying institutions, health supervision and inspection (check up) institutions, medicinal scientific research and on-job training institutions, health education centres and so on.

10. Medical Organizations include: hospitals, health service centres (stations) in communities, sanatoria, health centres, out-patient clinics, clinics (health stations and infirmaries), maternity and child care agencies (centres and stations), special disease prevention and curing agencies (centres and stations), first aid centres (stations) and clinical inspection centres. Medical organizations are grouped by two types: profit-making and non-profit-making medical organizations.

11. Energy Consumption per Unit of GDP refers to the energy consumption per unit of Gross Domestic Product in a country or the Gross Regional Product in a region in the same reference period. The formula is:

$$\text{Energy Consumption per Unit of GDP} = \frac{\text{Total Energy Consumption}}{\text{Gross Domestic Product}}$$

12. Electricity Consumption per Unit of GDP refers to the electricity consumption per unit of Gross Domestic Product in a country or the Gross Regional Product in a region in the same reference period. The formula is:

$$\text{Electricity Consumption per Unit of GDP} = \frac{\text{Total Electricity Consumption}}{\text{Gross Domestic Product}}$$

13. Energy Consumption per Unit of Industrial Value-added refers to the energy consumption per unit of industrial value-added in a country or region in the same reference period. The formula is:

$$\text{Energy Consumption per Unit of Industrial Value-added} = \frac{\text{Industry Energy Consumption}}{\text{Industrial Value-added.}}$$

14. Gross Amount of Water Used refers to gross water use distributed to users, including loss during transportation, broken down into use by agriculture, industry, living consumption and ecological protection.

15. Total Investment in Fixed Assets in the Whole Country refers to the volume of activities in construction and purchases of fixed assets of the whole country and related fees, expressed in monetary terms during the reference period. It is a comprehensive indicator which shows the size, structure and growth of the investment in fixed assets, providing a basis for observing the progress of construction projects and evaluating results of investment. Total investment in fixed assets in the whole country includes, by type of ownership, the investment by State-owned units, collective-owned units, individuals, joint ownership units, share-holding units, as well as investments by entrepreneurs from foreign countries and from Hong Kong, Macao and Taiwan, and by other units.

16. Total Imports and Exports at Customs refer to the real value of commodities imported and exported across the border of China. They include the actual imports and exports through foreign trade, imported and exported goods under the processing and assembling trades and materials, supplies and gifts as aid given gratis between governments and by the United Nations and other international organizations, and contributions donated by overseas Chinese compatriots in Hong Kong and Macao and Chinese with foreign citizenship, leasing commodities owned by tenant at the expiration of leasing period, the imported and exported commodities processed with imported materials, commodities trading in border areas (excluding mutual exchange goods), the imported and exported commodities and articles for public use of the Sino-foreign joint ventures, cooperative enterprises and ventures with sole foreign investment. Also included is the import or export of samples and advertising goods for which the CIF or FOB value is beyond the permitted ceiling (excluding goods of no trading or use value and free commodities for export), imported goods sold in China from bonded warehouses and other imported or exported goods. The indicator of the total imports and exports at customs can be used to observe the total size of external trade in a country. In accordance with the stipulation of the Chinese government, imports are calculated at CIF, while exports are calculated at FOB.

17. Import-Export Value by Location of China's Foreign Trade Managing Units refers to actual value of imports and exports carried out by corporations which have been registered by the local customs house and are vested with right to run import export business.

18. Disposable Income of Urban Households refers to the actual income at the disposal of members of the households which can be used for final consumption, other non-compulsory expenditure and savings. This equals to total income minus income tax, personal contribution to social security and subsidy for keeping diaries in being a sample household. The following formula is used:

Disposable income = total household income − income tax − personal contribution to social security - subsidy for keeping diaries for a sampled household

19. Consumption Expenditure of Urban Households refers to total expenditure of households for consumption in daily life, including expenditure on the eight categories of food; clothing; household appliances and services; health care and medical services; transport and communications; recreation, education and cultural services; housing; and miscellaneous goods and services.

20. Total Income refers to the sum of income earned from various sources by the rural households and their members during the reference period, and is classified as income from wages and salaries, household operations, properties and transfers.

21. Net Income refers to the total income of rural households from all sources minus all corresponding expenses. The formula for calculation is as follows:

Net income = total income − taxes and fees paid − household operation expenses − taxes and fees − depreciation of fixed assets for production − gifts to non-rural relatives

Net income is mainly used as input for reinvestment in production and as consumption expenditure of the year, and also used for savings and non-compulsory expenses of various forms. "Per capita net income of farmers" is the level of net income averaged by population, reflecting the average income level of rural households in a given area.

22. Total Population refers to the total number of people alive at a certain point of time within a given area.

The annual statistics on total population is taken at midnight, the 31st of December, not including residents in Taiwan province, Hong Kong and Macao and overseas Chinese.

23. Urban Population refers to all people residing in cities and towns.

24. Employed Persons refers to persons aged 16 and over who are engaged in gainful employment and thus receive remuneration payment or earn business income. This indicator reflects the actual utilization of total labour force during a certain period of time and is often used for the research on China's economic situation and national power.

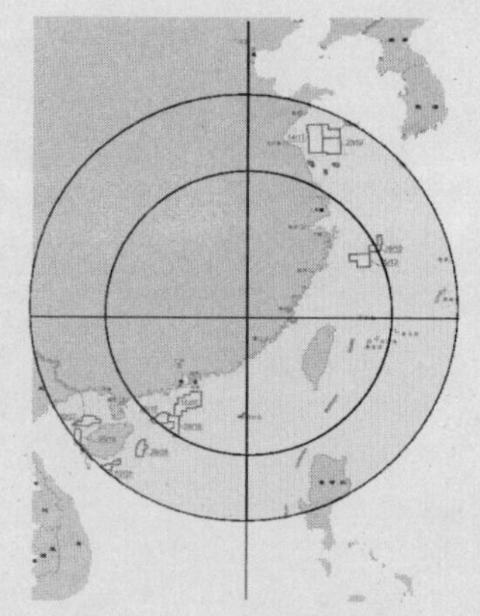

11

部分世界海洋经济统计资料

Part of the World's Marine Economic Statistics Data

11-1 世界海洋面积
World Ocean Area

区 域 Region	海洋面积（平方公里） Ocean Area (km^2)	占世界海洋面积的比重（%） Proportion in the World Ocean Area (%)	占地球表面面积的比重（%） Proportion in the Earth's Surface Area (%)
合 计 **Total**	**361 000 000**	**100.0**	**70.8**
太 平 洋 Pacific Ocean	178 334 000	49.4	35.0
大 西 洋 Atlantic Ocean	91 694 000	25.4	18.0
印 度 洋 Indian Ocean	76 171 000	21.1	14.9
北 冰 洋 Arctic Ocean	14 801 000	4.1	2.9

资料来源：《2011国际统计年鉴》。
Data Source: *International Statistical Yearbook 2011*.

11-2 主要沿海国家（地区）海岸线长度
Length of Coastline of Major Coastal Countries (Regions)

单位：千米 (km)

国家和地区 Country and Region	海岸线长度 Length of Coastline
美国 United States	22 680
日本 Japan	30 000
德国 Germany	1 300
英国 United Kingdom	11 450
法国 France	3 000
意大利 Italy	7 000
加拿大 Canada	20 000
澳大利亚 Australia	20 125
俄罗斯联邦 Russian Fed.	34 000
波兰 Poland	491
印度 India	6 083
印度尼西亚 Indonesia	35 000
菲律宾 Philippines	18 533
泰国 Thailand	3 219
马来西亚 Malaysia	4 675
新加坡 Singapore	193
缅甸 Myanmar	3 060
孟加拉国 Bangladesh	580
土耳其 Turkey	7 200
韩国 Korea, Rep.	2 413
埃及 Egypt	2 450
墨西哥 Mexico	9 330
巴西 Brazil	7 400
阿根廷 Argentina	4 989

11-3 世界主要沿海国家（地区）国土面积和人口（2011年）
Land Area and Population of Major Coastal Countries (Regions), 2011

国家和地区 Country and Region	国土面积(万平方千米) Area of Territory (10 000 km^2)	年中人口（万人） Mid-year Population (10 000 persons)	人口密度（人/平方千米） Population Density (people per km^2)
世界总计 World Total	**13 427.2**	**697 374**	**54**
中国 China	960.0	134 735	140
文莱 Brunei Darsm	0.6	41	77
柬埔寨 Cambodia	18.1	1 431	81
印度 India	328.7	124 149	418
印度尼西亚 Indonesia	190.5	24 233	134
日本 Japan	37.8	12 782	351
韩国 Korea, Rep	10.0	4 978	513
马来西亚 Malaysia	33.1	2 886	88
缅甸 Myanmar	67.7	4 834	74
菲律宾 Philippines	30.0	9 485	318
新加坡 Singapore	0.1	518	7 405
泰国 Thailand	51.3	6 952	136
越南 Viet Nam	33.1	8 784	283
埃及 Egypt	100.2	8 254	83
南非 South Africa	121.9	5 059	42
加拿大 Canada	998.5	3 448	4
墨西哥 Mexico	196.4	11 479	59
美国 United States	983.2	31 159	34
阿根廷 Argentina	278.0	4 076	15
巴西 Brazil	851.5	19 666	23
法国 France	54.9	6 543	119
德国 Germany	35.7	8 180	235
意大利 Italy	30.1	6 072	206
俄罗斯联邦 Russian Fed.	1 709.8	14 296	9
英国 United Kingdom	24.4	6 274	259
澳大利亚 Australia	774.1	2 232	3
新西兰 New Zealand	26.8	441	17

资料来源：《2013年中国统计年鉴》。

Data Source: *China Statistical Yearbook 2013*.

11-4 主要沿海国家（地区）国内生产总值
Gross Domestic Product of Major Coastal Countries (Regions)

国家和地区 Country and Region	2011年			2012年
	国内生产总值（亿美元） GDP (100 million USD)	人均国内生产总值（美元） GNI per Captia (current USD)	国内生产总值增长率(%) Growth Rate of GDP(%)	国内生产总值（亿美元） GDP (100 million USD)
中国 China	73 185	5 445	9.30	82 270
中国香港 Hong Kong,China	2 437	34 457	5.16	2 630
阿根廷 Argentina	4 460	10 941	8.87	4 750
澳大利亚 Australia	13 718	60 642	1.85	15 418
巴西 Brazil	24 767	12 594	2.73	23 960
柬埔寨 Cambodia	129	900	6.93	142
加拿大 Canada	17 361	50 345	2.46	18 191
埃及 Egypt	2 295	2 781	1.80	2 567
法国 France	27 730	42 377	1.70	26 087
德国 Germany	35 706	43 689	3.00	34 006
印度 India	18 480	1 489	6.86	18 248
印度尼西亚 Indonesia	8 468	3 495	6.46	8 782
意大利 Italy	21 948	36 116	0.43	20 141
日本 Japan	58 672	45 903	-0.70	59 640
韩国 Korea, Rep.	11 162	22 424	3.63	11 559
马来西亚 Malaysia	2 787	9 656	5.14	3 035
墨西哥 Mexico	11 553	10 064	3.94	11 771
菲律宾 Philippines	2 248	2 370	3.72	2 504
俄罗斯联邦 Russian Fed.	18 578	13 089	4.30	20 220
新加坡 Singapore	2 397	46 241	4.89	2 765
南非 South Africa	4 082	8 070	3.12	3 843
泰国 Thailand	3 456	4 972	0.05	3 656
英国 United Kingdom	24 316	38 818	0.66	24 405
美国 United States	150 940	48 442	1.70	156 848
越南 Viet Nam	1 240	1 411	5.89	1 381

资料来源：《2013年国际统计年鉴》。

Data Source: *International Statistical Yearbook 2013.*

11-5 世界主要沿海国家(地区)国内生产总值产业构成(2011年)
Industrial Composition of GDP of Major Coastal Countries (Regions) by Industry 2011

国家和地区 Country and Region	国内生产总值产业构成（%） Industrial Structure of GDP(%)		
	第一产业 Primary Industy	第二产业 Secondary Industry	第三产业 Tertiary Industy
世界 World	**2.8①**	**26.3①**	**70.9①**
中国 China	10.0	46.6	43.3
中国香港 Hong Kong,China	0.1③	7.4③	92.6③
文莱 Brunei Darsm	0.8①	66.8①	32.5①
柬埔寨 Cambodia	36.0①	23.3①	40.7①
印度 India	17.2	26.4	56.4
印度尼西亚 Indonesia	16.9	44.9	38.3
日本 Japan	1.2①	27.4①	71.5①
韩国 Korea, Rep.	2.6①	39.3①	58.2①
马来西亚 Malaysia	10.6①	44.4①	45.0①
缅甸 Myanmar	36.4①	26.0①	37.6①
菲律宾 Philippines	13.0	30.0	56.9
新加坡 Singapore		26.6	73.4
斯里兰卡 Sri Lanka	13.7	27.8	58.5
埃及 Egypt	13.9	36.7	49.3
南非 South Africa	2.4	30.6	67.0
泰国 Thailand	12.4	43.5	44.1
加拿大 Canada	1.9②	32.0②	66.1②
墨西哥 Mexico	3.7	34.1	62.2
美国 United States	1.2①	20.0①	78.8①
阿根廷 Argentina	9.1	26.3	64.6
巴西 Brazil	5.5	27.5	67.0
法国 France	1.8③	19.1③	79.2③
德国 Germany	0.9①	28.2①	71.0①
意大利 Italy	1.8①	25.2①	72.9①
荷兰 Netherlands	2.0①	23.9①	74.1①
俄罗斯联邦 Russian Fed.	4.0①	36.7①	59.3①
西班牙 Spain	2.7①	26.0①	71.3①
土耳其 Turkey	9.2	27.1	63.8
英国 United Kingdom	0.7①	21.7①	77.6①
澳大利亚 Australia	2.3①	19.8①	77.9①

注：①2010年数据。②2008年数据。③2009年数据。
Note: ①Data for 2010. ②Data for 2008. ③Data for 2009.

11-6 世界主要沿海国家（地区）就业人数
Employment in the Major Coastal Countries (Regions)

单位：万人　　　　(10 000 persons)

国家和地区 Country and Region	2000	2005	2007	2008	2009	2010	2011
中国 China	72 085	74 647	76 990	75 564	75 828	76 105	
中国香港 Hong Kong,China	321	334	319	353	350	352	
印度 India	33 020	37 199				37 429	
印度尼西亚 Indonesia	8 984	9 396	9 993	10 255	10 487	10 821	10 967
以色列 Israel	222	249	268	278	279	297	
日本 Japan	6 446	6 356	6 412	6 385	6 282	6 256	
韩国 Korea, Rep.	2 116	2 286	2 343	2 358	2 351	2 383	
马来西亚 Malaysia	932	1 005	1 102	1 066	1 090	1 113	
菲律宾 Philippines	2 745	3 231	3 356	3 409	3 500	3 604	3 719
新加坡 Singapore	209	165	180	185	187		
斯里兰卡 Sri Lanka	631	752	704	765	760	771	
泰国 Thailand	3 300	3 630	3 625	3 784	3 771	3 804	3 846
埃及 Egypt	1 720	1 934	2 153	2 251	2 278	2 383	
南非 South Africa	1 224	1 230	1 347	1 371	1 346	1 306	1 326
加拿大 Canada	1 476	1 617	1 681	1 713	1 681	1 704	
墨西哥 Mexico	3 950	4 079	4 306	4 387	4 291	4 387	
美国 United States	13 521	14 173	14 605	14 536	13 988	13 906	
阿根廷 Argentina	826	964	1 016	1 028	1 040	1 053	1 077
巴西 Brazil	6 563	8 719	2 044	9 239	9 269	2 202	
委内瑞拉 Venezuela	896	1 073	1 132	1 186	1 194	1 207	1 220
法国 France	2 312	2 495	2 558	2 590	2 565	2 569	2 576
德国 Germany	3 632	3 636	3 979	3 854	3 847	3 874	3 974
意大利 Italy	2 093	2 256	2 322	2 340	2 303	2 287	2 297
荷兰 Netherlands	786	811	731	859	860	837	837
波兰 Poland	1 453	1 412	1 524	1 580	1 587	1 596	1 613
俄罗斯联邦 Russian Fed.	6 507	6 817	7 055	7 097	6 700	6 979	
西班牙 Spain	1 544	1 897	2 036	2 026	1 889	1 846	1 810
土耳其 Turkey	2 158	2 205	2 079	2 119	2 127	2 259	2 410
乌克兰 Ukraine	2 127	2 068	2 091	2 097	2 019	2 027	
英国 United Kingdom	2 740	2 867	2 923	2 936	2 892	2 894	2 908
澳大利亚 Australia	895	997	1 058	1 074	1 092	1 125	
新西兰 New Zealand	181	207	217	219	216	219	

资料来源：《2013年国际统计年鉴》。

Data Source: *International Statistical Yearbook 2013.*

11-7 主要沿海国家（地区）渔业增加值
Fishery Added Value of Major Coastal Countries (Regions)

单位：亿本币 (100 million local currency units)

国家和地区 Country and Region	渔业增加值 Fishery Added Value						
	2000	2005	2006	2007	2008	2009	2010
孟加拉国 Bangladesh	1 367	1 546	1 630	1 780	1 979	2 181	2 422
文莱 Brunei Darsm	...	1		1	...	1	1
柬埔寨 Cambodia	15 159	18 922		24 300	31 163	33 015	35 491
印度 India	2 147	3 192	3 560	3 910	4 437	5 270	6 301
印度尼西亚* Indonesia	30	60	74	98	137	177	199
伊朗* Iran	2	4	5	5			
韩国* Korea, Rep.	2	2		2	2		
马来西亚 Malaysia	51	55	66	68	74	78	89
菲律宾 Philippines	785	1 160	1 300	1 430	1 700		
泰国 Thailand	1 184	1 061	1 090	1 030	968	1 163	1 275
越南* Viet Nam	15	33	38	46	58	62	74
尼日利亚 Nigeria	540	1 699		2 160	2 436	2 779	
土耳其 Turkey	4	17		22			
南非 South Africa	6	12			22	23	25
加拿大 Canada	11	13	13		15	15	
墨西哥 Mexico	64	73	62	62			
阿根廷 Argentina	6	16	20	18	24	22	25
委内瑞拉 Venezuela	2 200	9 500	9 880				
法国 France	14	15	14	14	13	11	
德国 Germany	2	2	2	2	2		
意大利 Italy	13	15			13		
荷兰 Netherlands	3	2	2	2	2		
波兰 Poland	2	2	2	2	1		
俄罗斯联邦 Russian Fed.		555		616	627	806	815
西班牙 Spain	15	16	16	17	17		
乌克兰 Ukraine	2	2	2	2	2	3	3
英国 United Kingdom	4	4					
澳大利亚 Australia	251	272	223	267	296		
新西兰 New Zealand	4		2				

注：*万亿本币。

Note: *Trillion local currency units.

11-8 主要沿海国家（地区）旅馆和饭店业增加值
Added Value of Hotel Industry in the Major Coastal Countries (Regions)

单位：亿本币 (100 million local currency units)

国家和地区 Country and Region	旅馆和饭店业增加值 Added Value of Hotel Industry						
	2000	2005	2006	2007	2008	2009	2010
中国 China	2 146	4 196	4 790	5 550	6 616	7 118	8 069
中国香港 Hong Kong, China	378	375	414	476	536	488	564
孟加拉国 Bangladesh	146	251	285	329	389	446	515
文莱 Brunei Darsm	1	1		1	1	1	1
柬埔寨 Cambodia	5 209	11 167		15 200	18 990	19 335	21 017
印度 India	2 513	5 432	6 610	7 790	8 289	8 908	10 577
印度尼西亚* Indonesia	39	93		124	140	158	179
伊朗* Iran	2	8	9	11			
韩国* Korea, Rep.	15	18	19	21	23		
马来西亚 Malaysia	80	117	124	142	176	192	214
菲律宾 Philippines	630	959	1 050	1 190	1 300		
新加坡 Singapore	35	41	43	50	60	55	63
斯里兰卡 Sri Lanka		142			206	250	332
泰国 Thailand	1 990	2 313	3 470	4 170	2 983	2 856	3 097
越南* Viet Nam	14	29	36	45	57	67	81
埃及 Egypt	66	177					
尼日利亚 Nigeria	65	461		728	865	994	
南非 South Africa		140					
加拿大 Canada	246	260	296		272	267	
墨西哥 Mexico	2 470	2 380	2 500	2 650			
美国 United States	2 830	3 643	3 580	3 800	4 074		
阿根廷 Argentina	78	126	163	200	251	273	335
巴西 Brazil	182	300		417	458	544	
委内瑞拉 Venezuela	1	6			26	36	44
法国 France	302	367	383	405	408	416	
德国 Germany	301	331	337	370	380		
意大利 Italy	416	483			541		
荷兰 Netherlands	76	85	88	93	90		
波兰 Poland	83	107	111	119	138		
俄罗斯联邦 Russian Fed.		1 678		2 860	3 580	3 437	3 886
西班牙 Spain	434	610	652	681	722		
乌克兰 Ukraine	8	24	54	68	97	80	103
英国 United Kingdom	260	327					
澳大利亚 Australia	183	238	256	269	285		
新西兰 New Zealand	21		31				

注：*万亿本币。
Note: *Trillion local currency units.

11-9 主要沿海国家（地区）运输、仓储和通讯业增加值
Added Values of Transportation,Storage and Communications Industries in the Major Coastal Countries (Regions)

单位：亿本币 (100 million local currency units)

国家和地区 Country and Region	运输、仓储和通讯业增加值 Added Values of Transportation, Storage and Commnuications Industries						
	2000	2005	2006	2007	2008	2009	2010
中国 China	6 161	10 666	12 200	14 600	16 363	16 727	19 132
中国香港 Hong Kong,China	1 400	1 601	1 370	1 420	1 465	1 459	1 921
孟加拉国 Bangladesh	1 974	3 829	4 320	4 890	5 691	6 428	7 188
文莱 Brunei Darsm	4	5		5	5	6	6
柬埔寨 Cambodia	9 301	19 038		24 200	31 020	32 236	35 653
印度 India	15 030	28 288	32 500	36 700	42 092	48 312	55 679
印度尼西亚* Indonesia	65	181	232	264	312	352	417
伊朗* Iran	48	148	191	240			
以色列 Israel	368	415	406	450	480	501	529
日本* Japan	35	34	34	34			
韩国* Korea, Rep	37	54	538 143	59	61		
马来西亚 Malaysia	249	364	389	426	465	469	511
巴基斯坦 Pakistan	4 010	7 600	9 080	10 100	11 800	16 100	
菲律宾 Philippines	1 990	4 140	4 460	4 780	5 090		
新加坡 Singapore	212	288	289	334	352	312	348
斯里兰卡 Sri Lanka	1 348	2 875	3 490	4 280	5 310	5 999	7 094
泰国 Thailand	4 179	5 668	5 690	6 260	7 013	7 111	7 404
越南* Viet Nam	17	37	44	51	66	72	85
土耳其 Turkey	203	891		1 170			
埃及 Egypt	257	574	580	767	924	1 011	1 104
尼日利亚 Nigeria	1 307	4 267		7 200	16 712	17 274	
南非 South Africa	809	1 395	1 540	1 640	1 898	1 993	2 037
加拿大 Canada	724	851	981		910	884	
墨西哥 Mexico	5 570	8 230	9 240	10 100			
美国 United States	6 215	7 442	7 720	8 190	8 133		
阿根廷 Argentina	241	444	536	641	803	867	1 073
巴西 Brazil	886	1 647	1 760	1 980	2 270	2 340	2 659
委内瑞拉* Venezuela	5	16			39	45	55
法国 France	776	998	1 030	1 080	1 117	1 131	
德国 Germany	1 018	1 162	1 220	1 230	1 280		
意大利 Italy	777	978			1 030		
荷兰 Netherlands	266	332	339	351	350		
波兰 Poland	434	627	685	722	783		
俄罗斯联邦 Russian Fed.	5 929	18 970	22 800	27 500	32 583	32 496	37 268
西班牙 Spain	418	563	601	641	663		
乌克兰 Ukraine	198	474	561	701	871	971	1 110
英国 United Kingdom	693	811					
澳大利亚 Australia	590	804	899	949	1 061		
新西兰 New Zealand	82		117				

注：*万亿本币。

Note: *Trillion local currency units.

11-10 渔获量居世界前25位的国家和地区（2010年—2011年）
World Top 25 Countries and Regions in Terms of Fish Catch (2010-2011)

单位：亿本币 (100 million local currency units)

国家和地区 Country and Region	2010	2011
世界总计 World Total	**88 970 124**	**93 494 340**
中国 China	15 417 011	15 772 054
秘鲁 Peru	4 261 091	8 248 482
印度尼西亚 Indonesia	5 380 196	5 707 684
美国 United States	4 425 961	5 153 452
印度 India	4 689 316	4 301 534
俄罗斯 Russian Fed.	4 069 624	4 254 864
日本 Japan	4 069 135	3 761 176
缅甸 Myanmar	3 063 210	3 332 979
智利 Chile	2 679 742	3 063 449
越南 Viet Nam	2 414 400	2 502 500
菲律宾 Philippines	2 611 762	2 363 221
挪威 Norway	2 680 187	2 281 429
泰国 Thailand	1 810 620	1 862 151
韩国 Korea, Rep.	1 733 310	1 746 998
孟加拉国 Bangladesh	1 726 586	1 600 918
墨西哥 Mexico	1 528 945	1 566 365
马来西亚 Malaysia	1 433 426	1 378 799
冰岛 Iceland	1 060 641	1 138 462
西班牙 Spain	971 511	993 457
摩洛哥 Morocco	1 136 240	958 907
中国台湾 Taiwan China	851 384	903 831
加拿大 Canada	936 090	861 388
巴西 Brazil	785 369	803 267
阿根廷 Argentina	881 749	792 505
丹麦 Denmark	828 016	716 312

11-11 食用鱼类养殖产量居世界前20位国家和地区（2010年—2011年）

World Top 20 Aquaculture Producers of Food Fish in 2010 and 2011

单位：吨 (t)

2010		2011	
国家和地区 Country and Region	产 量 Quantity	国家和地区 Country and Region	产 量 Quantity
世界总计 World Total	**59 022 185**	**世界总计 World Total**	**62 700 300**
20生产国合计 Total of Top-20 Producers	55 917 410	**20生产国合计 Total of Top-20 Producers**	59 474 657
中国 China	36 734 215	中国 China	38 621 269
印度 India	3 785 779	印度 India	4 573 465
越南 Viet Nam	2 671 800	越南 Viet Nam	2 845 600
印度尼西亚 Indonesia	2 304 828	印度尼西亚 Indonesia	2 718 421
孟加拉国 Bangladesh	1 308 515	孟加拉国 Bangladesh	1 523 759
泰国 Thailand	1 286 122	挪威 Norway	1 138 797
挪威 Norway	1 008 010	泰国 Thailand	1 008 049
埃及 Egypt	919 585	埃及 Egypt	986 820
缅甸 Myanmar	850 697	智利 Chile	954 845
菲律宾 Philippines	744 695	缅甸 Myanmar	816 820
日本 Japan	718 284	菲律宾 Philippines	767 287
智利 Chile	701 062	巴西 Brazil	629 309
美国 United States	496 699	日本 Japan	556 761
巴西 Brazil	479 399	韩国 Korea, Rep.	507 052
韩国 Korea, Rep.	475 561	美国 United States	396 841
马来西亚 Malaysia	373 151	中国台湾 China, Taiwan	314 363
中国台湾 China, Taiwan	310 338	厄瓜多尔 Ecuador	308 900
厄瓜多尔 Ecuador	271 919	马来西亚 Malaysia	287 076
西班牙 Spain	252 351	西班牙 Spain	271 961
法国 France	224 400	伊朗 Iran	247 262
其他 Other	3 104 775	其他 Other	3 225 644

注：食用鱼类是指人类消费食用的鱼类、甲壳类、软件动物类、两栖类、爬行类（鳄鱼除外）及其他的水生动物（如海参、海胆等）。

Note: *Food Fishes=fishes, crustaceans, molluscs, amphibians, reptiles (excluding crocodiles) and other aquatic animals (such as sea cucumber, sea urchin, etc.) for human consumption.

11-12 主要沿海国家（地区）油气产量（2011年）
Natural Gas and Crude Oil Production of the Major Coastal Countries (Regions), 2011

单位：万标准油 (10 000 TOE)

国家和地区 Country and Region	原油产量 Production of Crude Oil	天然气产量 Production of Natural Gas
中国 China	20 320	8 626
孟加拉国 Bangladesh	6	1 649
文莱 Brunei Darsm	786	1 071
印度 India	4 421	3 837
印度尼西亚 Indonesia	4 660	7 997
伊朗 Iran	21 925	12 598
以色列 Israel		352
日本 Japan	67	320
韩国 Korea, Rep.	69	41
马来西亚 Malaysia	3 139	4 612
缅甸 Myanmar	89	994
巴基斯坦 Pakistan	348	2 657
菲律宾 Philippines	72	307
泰国 Thailand	1 638	2 335
越南 Viet Nam	1 525	739
埃及 Egypt	3 566	4 980
尼日利亚 Nigeria	14 168	2 933
南非 South Africa	7	49
加拿大 Canada	17 274	13 280
墨西哥 Mexico	15 953	4 219
美国 United States	36 123	53 455
阿根廷 Argentina	3 298	3 419
巴西 Brazil	11 248	1 417
委内瑞拉 Venezuela	15 801	2 384
法国 France	109	51
德国 Germany	346	900
意大利 Italy	562	692
荷兰 Netherlands	173	5 773
波兰 Poland	68	385
俄罗斯联邦 Russian Fed.	51 265	55 624
西班牙 Spain	10	5
土耳其 Turkey	234	63
乌克兰 Ukraine	340	1 552
英国 United Kingdom	5 328	4 070
澳大利亚 Australia	2 046	4 950
新西兰 New Zealand	231	348

资料来源：《2013年国际统计年鉴》。
Data Source: *International Statistical Yearbook 2013.*

11-13 主要沿海国家（地区）能源利用效率
Energy Efficiency in the Major Coastal Countries (Regions)

单位：吨标准油/万美元 (ton of oil equivalent per 10 000 USD)

国家和地区 Country and Region	单位国内生产总值能耗 Energy Consumption Per Unit GDP						
	2000	2004	2005	2006	2007	2008	2009
世界 World	**3.01**	**3.04**	**3.00**	**2.96**	**2.92**	**2.93**	**2.97**
中国 China	9.14	9.12	8.89	8.61	7.99	7.86	7.68
中国香港 Hong Kong,China	0.79	0.66	0.61	0.60	0.61	0.59	0.64
孟加拉国 Bangladesh	3.95	3.89	3.89	3.88	3.80	3.78	3.78
文莱 Brunei Darsm	4.09	3.92	3.82	4.69	4.78	5.32	4.66
柬埔寨 Cambodia	10.89	9.21	8.33	7.84	7.32	6.98	6.96
印度 India	9.63	8.80	8.17	8.00	7.55	7.54	7.61
印度尼西亚 Indonesia	9.43	8.94	8.72	8.27	8.07	7.76	7.81
伊朗 Iran	12.76	12.61	13.24	12.74	12.88	13.33	13.65
以色列 Israel	1.46	1.49	1.46	1.42	1.39	1.39	1.33
日本 Japan	1.10	1.07	1.04	1.02	0.99	0.96	0.97
韩国 Korea, Rep.	3.53	3.26	3.16	3.06	3.02	3.02	3.04
马来西亚 Malaysia	5.04	4.86	5.29	5.06	5.22	5.23	4.86
巴基斯坦 Pakistan	8.59	8.43	8.07	7.95	7.95	7.70	7.67
菲律宾 Philippines	4.99	4.10	3.86	3.69	3.35	3.34	3.24
新加坡 Singapore	2.01	1.75	1.54	1.79	1.09	1.15	1.28
斯里兰卡 Sri Lanka	5.10	4.71	4.54	4.25	4.06	3.71	3.71
泰国 Thailand	5.90	6.27	6.10	6.04	5.88	5.97	5.94
越南 Viet Nam	11.84	12.09	11.38	10.88	10.66	10.59	10.88
埃及 Egypt	4.52	4.92	5.12	5.03	4.95	4.86	4.72
尼日利亚 Nigeria	19.53	17.32	16.83	16.00	15.13	14.95	13.64
南非 South Africa	8.61	8.44	8.13	7.61	7.77	8.10	7.90
加拿大 Canada	3.47	3.35	3.31	3.18	3.15	3.06	3.01
墨西哥 Mexico	2.50	2.58	2.67	2.56	2.54	2.59	2.66
美国 United States	2.30	2.13	2.08	2.01	2.00	1.96	1.93
阿根廷 Argentina	2.14	2.34	2.14	2.15	1.99	1.94	1.87
巴西 Brazil	2.93	2.93	2.91	2.90	2.89	2.90	2.81
委内瑞拉 Venezuela	4.82	4.68	5.01	4.17	4.04	3.94	4.13
法国 France	1.90	1.91	1.88	1.81	1.76	1.78	1.76
德国 Germany	1.79	1.77	1.74	1.69	1.59	1.59	1.60
意大利 Italy	1.55	1.59	1.59	1.56	1.49	1.48	1.46
荷兰 Netherlands	1.90	1.96	1.92	1.81	1.80	1.77	1.80
波兰 Poland	5.20	4.75	4.63	4.59	4.28	4.12	3.89
俄罗斯联邦 Russian Fed.	23.84	19.69	18.64	17.73	16.38	15.94	16.25
西班牙 Spain	2.10	2.11	2.08	1.99	1.96	1.87	1.78
土耳其 Turkey	2.86	2.63	2.53	2.61	2.68	2.63	2.74
乌克兰 Ukraine	42.80	32.67	31.59	28.30	26.23	25.40	25.30
英国 United Kingdom	1.51	1.36	1.31	1.27	1.16	1.16	1.15
澳大利亚 Australia	2.59	2.40	2.44	2.43	2.39	2.38	2.38
新西兰 New Zealand	3.26	2.75	2.64	2.61	2.60	2.69	2.73

资料来源：《2013年国际统计年鉴》。

Data Source: *International Statistical Yearbook 2013.*

11-14 主要沿海国家（地区）国际海运装货量和卸货量

International Ocean Shipping Loading and Unloading Capacities in the Major Coastal Countries (Regions)

单位：万吨 (10 000 tons)

国家和地区 Country and Region	国际海运装货量 International Ocean Shipping Loading Capacity			国际海运卸货量 International Ocean Shipping Unloading Capacity		
	2000	2005	2011	2000	2005	2011
中国香港 Hong Kong,China	6 770	8 918	11 906	10 693	14 095	15 714
孟加拉国 Bangladesh	89	79	466 ①	1 408		3 860 ①
文莱 Brunei Darsm	10	8		102	168	
印度尼西亚 Indonesia	14 153	27 372	50 118 ①	4 504	8 479	11 254 ①
伊朗 Iran	3 065	3 446	4 366 ②	4 486	6 073	8 232 ②
以色列 Israel	1 387	1 663	1 942	2 920	2 107	2 510
日本 Japan	13 010			80 654		
韩国 Korea Rep.	15 078	24 250		41 882	51 245	
马来西亚 Malaysia	5 483	7 940	12 547	6 922	10 391	15 305
巴基斯坦 Pakistan	617	1 126	1 950 ①	3 080	3 955	4 870 ①
新加坡③ Singapore	32 618	42 266	47 146 ②			
斯里兰卡 Sri Lanka	919	1 315				
埃及 Egypt		2 123			4 241	
南非 South Africa	2 598					
美国 United States	34 334	40 534		83 352	94 160	
阿根廷④ Argentina	1 550					2 374 ①
法国 France	6 810	10 066	9 986 ①	20 273	22 710	19 931 ①
德国 Germany	8 602	10 832	11 248	14 725	16 866	17 706
荷兰 Netherlands	9 940	12 250	15 335 ②	32 507	36 424	35 617 ②
波兰 Poland	3 152	3 835	2 394	1 582	1 642	3 264
俄罗斯联邦 Russian Fed.	828	910	18 924	84	74	2 184
西班牙 Spain	5 627	8 195		19 343	26 164	
乌克兰 Ukraine	4 271	7 070	8 490	684	1 333	1 943
澳大利亚 Australia	48 750	62 401	91 271	5 418	6 989	9 346
新西兰 New Zealand	2 214	2 167	3 174	1 379	1 844	1 892

注：①2010年数据。②2009年数据。③包括装货量和卸货量。④不包括转口。

Note: ①Data for 2010. ②Data for 2009. ③ Including both goods loaded and unloaded.④ fransit goods loaded and

资料来源：《2013年国际统计年鉴》。

Data Source: *International Statistical Yearbook 2013.*

11-15 世界主要外贸货物海运量及构成*（2011年）
World Major Maritime Freight Traffic in Foreign Trade, 2011

品种 Sort	海运量（百万吨） Freight Traffic (million tons)	
	2011	所占比例（%） Percentage
合计 Total	**8 839**	**100**
原油 Crude Oil	2 016	23
成品油 Refined Oil	822	9
燃气 Fuel gas	234	3
铁矿石 Ironstone	1 060	12
煤炭 Coal	954	11
谷物 Corn	355	4
其他货物 Others	3 396	38

注：*为估计数。
资料来源：Shipping Statistics and Market Review, January/February 2011, ISL
Note: *Estimated data.
Data Source: *Shipping Statistics and Market Review*, January/February 2011, ISL.

11-16 主要沿海国家（地区）国际旅游人数
Number of International Tourists of Major Coastal Countries (Regions)

单位：万人 (10 000 persons)

国家和地区 Country and Region	入境（过夜）旅游人数 Number of Inbound Tourists (Overnight)			出境旅游人数 Number of Outbound Tourists		
	2000	2005	2010	2000	2005	2010
世界总计 World Total	**68 113**	**80 493**	**94 164**	**73 089**	**85 387**	
中国 China	3 123	4 681	5 566	1 047	3 103	5 739
中国香港 Hong Kong, China	881	1 477	2 009	461	7 230	8 444
中国澳门 Macao, China	520	901	1 193	14	30	25
孟加拉国 Bangladesh	20	21		113	177	
文莱 Brunei Darsm	98	13				
柬埔寨 Cambodia		133	240	4	57	51
印度 India	265	392	578	442	719	1 299
印度尼西亚 Indonesia	506	500	700	221	411	624
伊朗 Iran	134	189		229		
以色列 Israel	242	190	280	353	369	427
日本 Japan	476	673	861	1 782	1 740	1 664
韩国 Korea, Rep.	532	602	880	551	1 008	
马来西亚 Malaysia	1 022	1 643	2 458	3 053		
缅甸 Myanmar	21	23	31			
巴基斯坦 Pakistan	56	80				
菲律宾 Philippines	199	262	352	167	214	
新加坡 Singapore	606	708	916	444	516	734
斯里兰卡 Sri Lanka	40	55	65	52	73	112

11-16 续表 continued

国家和地区 Country and Region	入境（过夜）旅游人数 Number of Inbound Tourists (Overnight)			出境旅游人数 Number of Outbound Tourists		
	2000	2005	2010	2000	2005	2010
泰国 Thailand	958	1 157	1 594	191	305	545
越南 Viet Nam	214	348				
埃及 Egypt	512	824	1 405	296	531	
尼日利亚 Nigeria	81	101				
南非 South Africa	587	737	807	383		517
加拿大 Canada	1 963	1 877	1 610	1 918	2 110	2 868
墨西哥 Mexico	2 064	2 192	2 226	1 108	1 331	1 440
美国 United States	5 124	4 921	5 979	6 133	6 350	
阿根廷 Argentina	291	382	533	495	389	531
巴西 Brazil	531	536	516	323	347	531
委内瑞拉 Venezuela	47	71		95	107	
法国 France	7 719	7 499	7 715	1 989	2 248	2 161
德国 Germany	1 898	2 150	2 688	7 440	7 740	
意大利 Italy	4 118	3 651	4 363	2 199	2 480	2 982
荷兰 Netherlands	1 000	1 001	1 088	1 390	1 704	1 843
波兰 Poland	1 740	1 520	1 247	5 668	4 084	4 276
俄罗斯联邦 Russian Fed.	2 117	2 220	2 228	1 837	2 842	3 932
西班牙 Spain	4 640	5 591	5 268	410	1 046	
土耳其 Turkey	959	2 027	2 700	528	825	1 100
乌克兰 Ukraine	643	1 763	2 120	1 342	1 645	1 718
英国 United Kingdom	2 321	2 804	2 830	5 684	6 649	5 556
澳大利亚 Australia	493	550	589	350	476	711
新西兰 New Zealand	178	235	249	128	187	203

资料来源：世界银行WDI数据库。

Data source: *WDI database of World Bank.*

11-17 主要沿海国家（地区）国际旅游收入
International Tourism Receipts of Major Coastal Countries (Regions)

单位：亿美元　　(10 000 million USD)

国家和地区 Country and Region	国际旅游收入 International Tourism Receipts		
	2009	2010	2011
世界总计　World Total	**10 306**	**11 201**	**11 562**
中国 China	426	502	533
中国香港　Hong Kong, China	203	272	337
中国澳门　Macao, China	184	282	
孟加拉国 Bangladesh	1	1	1
文莱 Brunei Darsm	3		
柬埔寨 Cambodia	12	13	18
印度 India	111	142	175
印度尼西亚 Indonesia	61	76	90
伊朗 Iran	23	31	
以色列 Israel	43	55	56
日本 Japan	125	154	125
韩国 Korea, Rep.	133	138	172
马来西亚 Malaysia	172	183	196
缅甸 Myanmar	1	1	3
巴基斯坦 Pakistan	10	10	11
菲律宾 Philippines	29	32	38
新加坡 Singapore	94	141	180
斯里兰卡 Sri Lanka	8	10	14

11-17 续表 continued

国家和地区 Country and Region	国际旅游收入 International Tourism Receipts		
	2009	2010	2011
泰国 Thailand	198	238	309
越南 Viet Nam	31	45	56
埃及 Egypt	118	136	93
尼日利亚 Nigeria	8	7	7
南非 South Africa	87	103	107
加拿大 Canada	156	183	199
墨西哥 Mexico	125	126	123
美国 United States	1 495	1 651	1 859
阿根廷 Argentina	45	56	61
巴西 Brazil	56	62	68
委内瑞拉 Venezuela	11	8	8
法国 France	589	563	652
德国 Germany	475	491	534
意大利 Italy	419	401	454
荷兰 Netherlands	179	187	210
波兰 Poland	98	100	116
俄罗斯联邦 Russian Fed.	124	132	170
西班牙 Spain	597	590	675
土耳其 Turkey	246	248	281
乌克兰 Ukraine	43	47	54
英国 United Kingdom	386	407	459
澳大利亚 Australia	280	323	342
新西兰 New Zealand	46	49	55

资料来源：《2013年中国统计年鉴》。

Data Source: *China Statistical Yearbook 2013*.

11-18 集装箱吞吐量居世界前20位的港口
World Top 20 Seaports in Terms of the Number of Containers Handled

单位：万标准箱 (10 000 TEU)

港　口 Seaport	所属国家或地区 Country or Region	吞吐量 Containers Handled
上海 Shanghai	中国 China	3 253
新加坡 Singapore	新加坡 Singapore	3 165
香港 Hong Kong	中国 China	2 313
深圳 Shenzhen	中国 China	2 294
釜山 Pusan	韩国 Korea, Rep.	1 704
宁波-舟山 Ningbo and Zhoushan	中国 China	1 617
广州 Guangzhou	中国 China	1 455
青岛 Qingdao	中国 China	1 450
迪拜 Dubayy	阿联酋 United Arab Em	1 330
天津 Tianjin	中国 China	1 230
鹿特丹 Rotterdam	荷兰 Netherlands	1 190
巴生 Kelang	马来西亚 Malaysia	1 001
高雄 Gaoxiong	中国台湾 Taiwan, China	978
汉堡 Hamburg	德国 Germany	886
安特卫普 Antwerp	比利时 Belgium	864
洛杉矶 Los Angeles	美国 United States	808
大连 Dalian	中国 China	806
丹戎帕拉帕斯 Tanjung Periuk	马来西亚 Malaysia	772
厦门 Xiamen	中国 China	720
长滩 Long Beach	美国 United States	605

资料来源：上海航运交易所，CL-ONLINE。
Data Source: *Shanghai Shipping Exchange, CL-ONLINE.*

11-19　港口货物吞吐量居世界前20位的港口（2011年）
World Top 20 Seaports in Terms of the Cargo Handled, 2011

单位：百万吨　　(million tons)

港　口 Seaport	所属国家或地区 Country or Region	吞吐量 Cargo Handled	备　注 Remark
上海 Shanghai	中国 China	727.6	内外贸货物 Domestic and Foreign Trade Cargo
新加坡 Singapore	新加坡 Singapore	531.2	内外贸货物 Domestic and Foreign Trade Cargo
天津 Tianjin	中国 China	453.4	内外贸货物 Domestic and Foreign Trade Cargo
鹿特丹 Rotterdam	荷兰 Netherlands	434.6	外贸货物 Foreign Trade Cargo
宁波 Ningbo	中国 China	433.4	内外贸货物 Domestic and Foreign Trade Cargo
广州 Guangzhou	中国 China	431.5	内外贸货物 Domestic and Foreign Trade Cargo
青岛 Qingdao	中国 China	372.3	内外贸货物 Domestic and Foreign Trade Cargo
大连 Dalian	中国 China	336.9	内外贸货物 Domestic and Foreign Trade Cargo
秦皇岛 Qinhuangdao	中国 China	287.7	内外贸货物 Domestic and Foreign Trade Cargo
釜山 Pusan	韩国 Korea, Rep.	281.5	内外贸货物 Domestic and Foreign Trade Cargo
香港 Hong Kong	中国 China	277.4	内外贸货物 Domestic and Foreign Trade Cargo
休斯敦 Houston	美国 United States	261.7	外贸货物 Foreign Trade Cargo
南路易斯安娜 Southern Louisiana	美国 United States	248.8	内外贸货物 Domestic and Foreign Trade Cargo
黑德兰港 Headland Harbour	澳大利亚 Australia	246.7	内外贸货物 Domestic and Foreign Trade Cargo
深圳 Shenzhen	中国 China	223.3	内外贸货物 Domestic and Foreign Trade Cargo
巴生 Kelang	马来西亚 Malaysia	193.7	内外贸货物 Domestic and Foreign Trade Cargo
安特卫普 Antwerp	比利时 Belgium	187.2	内外贸货物 Domestic and Foreign Trade Cargo
名古屋 Nagoya	日本 Japan	186.3	内外贸货物 Domestic and Foreign Trade Cargo
长滩 Long Beach	美国 United States	185.6	内外贸货物 Domestic and Foreign Trade Cargo
丹皮尔 Dampier	澳大利亚 Australia	167.2	内外贸货物 Domestic and Foreign Trade Cargo

资料来源：Shipping Statistics and Market Review, December 2012, ISL.

Data Source: *Shipping Statistics and Market Review, December 2012, ISL.*

11-20 海上商船拥有量居世界前20位的国家或地区
World Top 20 Countries or Regions in Terms of the Number of Maritime Merchant Ships Owned

国家和地区 Country and Region	艘数（艘） Number of Vessels (unit)	载重吨 Deadweight ton	
		万吨 10 000 tons	占世界% Percentage in the World
世界总计 World Total	**48 197**	**146 175.9**	**100.0**
巴拿马 Panama	6 741	31 970.4	21.9
利比里亚 Liberia	2 865	18 287.7	12.5
马绍尔群岛 Marshall Islamds	1 677	11 586.4	7.9
中国香港 Hong Kong, China	1 805	11 280.1	7.7
新加坡 Singapore	1 765	7 888.8	5.4
希腊 Greece	1 074	7 246.2	5.0
马耳他 Malta	1 693	6 984.8	4.8
巴哈马 Bahamas	1 182	6 277.3	4.3
中国 China	2 590	5 613.1	3.8
英国 United Kingdom	908	3 964.1	2.7
塞浦路斯 Cyprus	842	3 279.9	2.2
日本 Japan	2 367	2 318.8	1.6
意大利 Italy	824	2 127.1	1.5
韩国 Korea, Rep.	1 032	1 884.3	1.3
挪威 Norway	845	1 800.1	1.2
德国 Germany	449	1 760.4	1.2
印度 India	456	1 536.1	1.1
安提瓜和巴布亚 Antigua and Papua	1 270	1 434.2	1.0
丹麦 Denmark	447	1 370.8	0.9
印度尼西亚 Indonesia	2 475	1 282.7	0.9

注：1. 表中数据按载重吨排序；

2. 统计范围为300总吨及以上船舶，截至日期均为2012年1月1日；

3. 因统计口径不同，表中数据与其他出版物公布的数据略有差别。

资料来源：Shipping statistics and Market Review, Jannuang/February 2012. ISL.

Note: 1. The data in the table are arranged in the order of deadweight tons.

2. The ships, ranging over 300 tons and more in gross ton, are counted as of Jan. 1, 2012.

3. Owing to different statistical requirements, the data given in the table may be somewhat different from those issued in other publications.

Data Source: *Shipping statistics and Market Review, Jannuang/February 2012. ISL.*

11-21 集装箱船拥有量居世界前20位的国家或地区
World Top 20 Countries or Regions in Terms of the Number of Container Ships Owned

国家和地区 Country and Region	艘数（艘）Number of Vessels (unit)	载重吨 Deadweight ton	
		万标准箱 10 000 TEU	占世界% Percentage in the World
世界总计 World Total	**4 570**	**1 469.8**	
利比里亚 Liberia	975	341.7	22.3
巴拿马 Panama	738	291.7	19.1
德国 Germany	285	123.1	8.0
中国香港 Hong Kong, China	290	118.0	7.7
新加坡 Singapore	343	94.8	6.2
英国 United Kingdom	182	86.0	5.6
马绍尔群岛 Marshall Islamds	228	65.4	4.3
丹麦 Denmark	97	60.0	3.9
安提瓜和巴布亚 Antigua and Papua	405	57.0	3.7
中国 China	219	45.0	2.9
马耳他 Malta	118	38.9	2.5
塞浦路斯 Cyprus	196	36.9	2.4
美国 United States	84	27.1	1.8
希腊 Greece	35	19.7	1.3
法国 France	26	17.6	1.1
巴哈马 Bahamas	62	15.7	1.0
荷兰 Netherlands	67	9.9	0.6
意大利 Italy	21	7.9	0.5
韩国 Korea, Rep.	73	6.8	0.4
印度尼西亚 Indonesia	126	6.6	0.4

注：1. 表中数据按集装箱位排序；

2. 统计范围为300总吨及以上船舶，截至日期均为2012年1月1日；

3. 因统计口径不同，表中数据与其他出版物公布的数据略有差别。

资料来源：Shipping statistics and Market Review, Jannuang/February 2012. ISL.

Note: 1. The data in the table are arranged in the order of deadweight tons.

2. The ships, ranging over 300 tons and more in gross ton, are counted as of Jan. 1, 2012.

3. Owing to different statistical requirements, the data given in the table may be somewhat different from those issued in other publications.

Data Source: *Shipping statistics and Market Review, Jannuang/February 2012. ISL.*

图书在版编目（CIP）数据

中国海洋统计年鉴．2013/国家海洋局编著．--北京：海洋出版社，2014.4

ISBN 978-7-5027-8864-3

Ⅰ．①中… Ⅱ．①国… Ⅲ．①海洋—统计资料—中国—2013—年鉴 Ⅳ．①P7-66

中国版本图书馆 CIP 数据核字（2014）第 078407 号

责任编辑：王　溪

责任印制：赵麟苏

海洋出版社 出版发行

http: //www.oceanpress.com.cn

北京市海淀区大慧寺路 8 号　　　邮编：100081

国家海洋信息中心印刷厂印刷　　　新华书店发行所经销

2014 年 4 月第 1 版　　2014 年 4 月第 1 次印刷

开本：787mm×1092mm　1/16　印张：19.75

字数：495 千字

定价：168.00 元

发行部：(010)62147016　　邮购部：(010)68038093　　总编室：(010)62114335

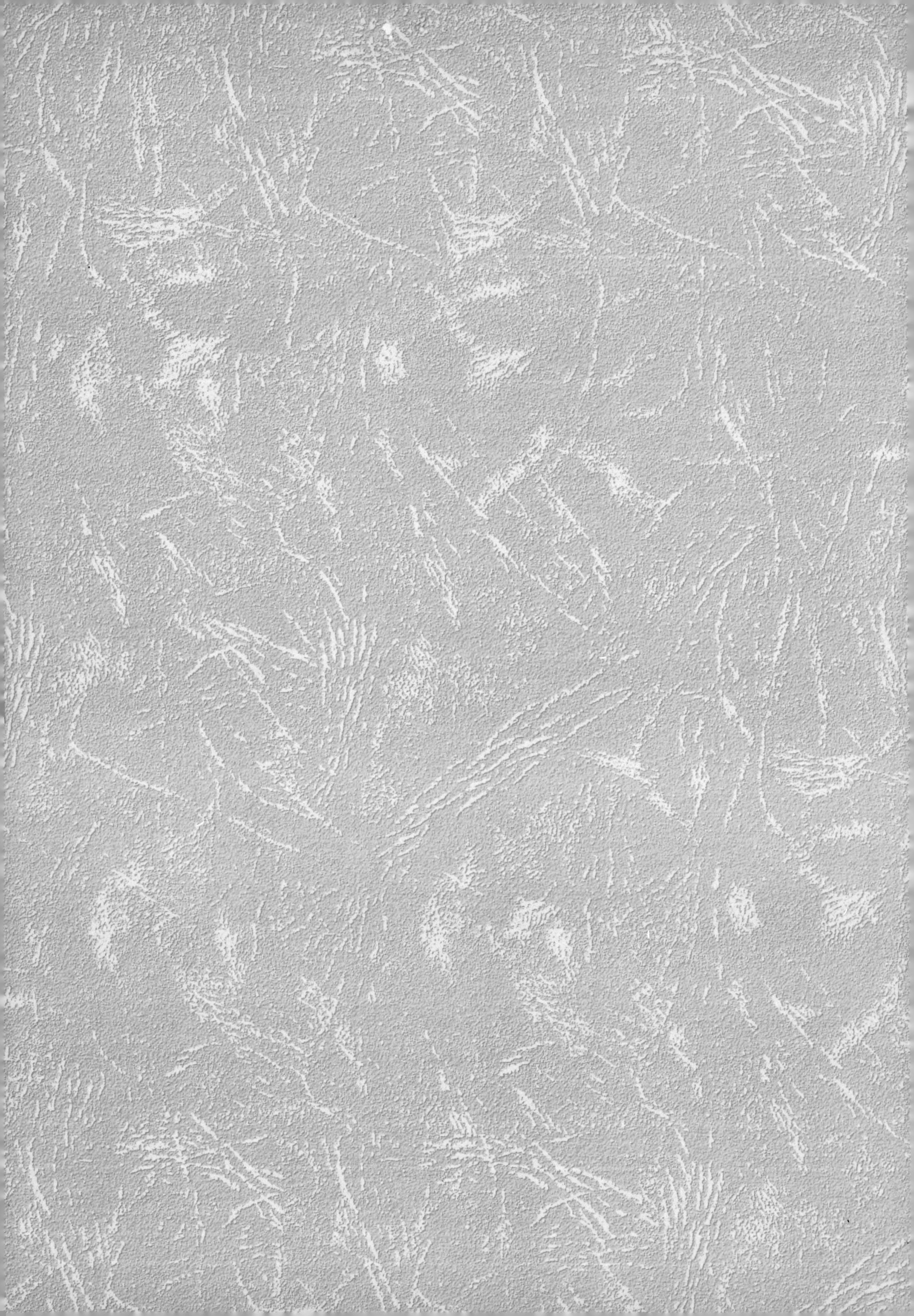